Growth and Development

Robert E. Ulanowicz

Growth and Development
Ecosystems Phenomenology

With 54 Figures

toExcel
San Jose New York Linclon Shanghai

Growth and Development
Ecosystems Phenomenology

This edition published by toExcel Press,
an imprint of iUniverse.com, Inc.

For information address:
iUniverse.com, Inc.
620 North 48th Street
Suite 201
Lincoln, NE 68504-3467
www.iuniverse.com

ISBN: 0-595-00145-9

To Marijka

omnibus formosior,
semper in te glorior!

Preface

"What in the ever-loving blue-eyed world do these [Ulano-wicz's] innocuous comments on thermodynamics have to do with ecology!"

Anonymous manuscript reviewer
The American Naturalist, 1979

"The germ of the idea grows very slowly into something recognizable. It may all start with the mere desire to have an idea in the first place."

Walt Kelly
Ten Ever-Lovin' Blue-Eyed Years with Pogo, 1959

"It all seems extremely interesting, but for the life of me it sounds as if you pulled it out of the air," my good friend Ray Lassiter exclaimed to me after enduring about 20 minutes of my enthusiasm for the newly formulated concept of "ascendency" in ecosystems. "It wasn't," I replied, "but it would take a book to show you where it came from."

If such was the reaction of someone usually sympathetic to my manner of thinking, what could I expect from those who viewed biological development in the traditional way? After all, I was suggesting that it is possible to quantify the growth and development of an entire ecosystem. Furthermore, I was maintaining that this development was not entirely determined by events and entities at smaller scales, and yet could influence these component processes and structures.

To be sure, mine was only the latest of many challenges to straight reductionism, but, like everyone else with a new idea, I thought mine was special. It was formulated in terms that evolved quite naturally from widely accepted phenomenological principles. Practically every other opposition to strict reductionism had generated heated controversy, so I was anxious for reaction to my synthesis—be it laudatory or vituperative! But response was slow in coming.

From some, it was a bemused "too transcendental" or "dream-like." From a very few, there came the expected fulminations. But largely, I was met with indifference. I spent innumerable hours in imaginary confrontations with critics to prepare myself for entering reputed strongholds of neo-Darwinism. With few exceptions, those preparations proved unnecessary, as question periods were generally short, and most potential critics simply filed out of the seminar room without having uttered a word. Despite my obvious enthusiasm, it became clear that I was not communicating the thrust of my initiative.

The source of much of the indifference to my thoughts is apparent to me now. My formal education was in chemical engineering *science.* During the years I was in graduate school, my institution was deemphasizing the strictly vocational aspects of engineering and stressing instead those elements of science that supported the profession. Much time, therefore, was devoted to evaluating the motivations and perspectives of those who significantly advanced these underlying disciplines. Thermodynamics was a particularly fertile domain for discussion, and I believe that most of us left the department with a high regard for pragmatic quantitative methods, phenomenological inquiry, and the macroscopic approach to quantifying whole systems.

By contrast, most of my audiences in biology departments had spent their formative years in growing admiration for the analytical process—dissecting that which is whole, probing that which is small, and looking for causes in component parts. Not only was there substantial disparity in the subjects we had studied, but factual matter had been assimilated under divergent perspectives. Perhaps what I was proposing was too foreign even to excite my colleagues' ire. Clearly, if I wanted to be taken seriously, I needed to write the book I had first mentioned to Ray.

Thus, it is with my coworkers in biology foremost in mind that I have written this text. My aim is to introduce everyone to the conceptual elements that gave rise to my theory without unduly burdening the reader with higher mathematics, technical details, or esoteric jargon. Whereas the reader may begin the introduction feeling that it is nonsense to talk about the growth and development of an ecosystem, I hope that by the sixth chapter the notion will no longer seem arbitrary or strange. True, to accommodate the phenomenological perspective will require many biologists to adjust some of their popularly held attitudes, but the ideas used in this synthesis are not so new or radical as to require the reader to renounce the entire framework upon which his career is built!

I would encourage potential readers who are not biologists to take heart. The subject of growth and development is broad enough never to allow this exposition to degenerate to where it has little interest to those outside biology. Scattered copiously throughout the chapters are implications and ramifications certain to pique the interest of those in fields such as economics, thermodynamics, cybernetics, cognitive science, network analysis, operations research, fluid mechanics, sociology, and possibly even philosophy.

Having emphasized the disparity between the backgrounds of the engineer and the biologist as a barrier to communication, I should confess that some other impediments to useful exchange were of my own unintentional doing. As I look back at my early publications on network development, they now appear to me rather inchoate and cryptic. Given that they were probably more coherent than my verbal presentations, it is little wonder that many readers and listeners seemed mystified, or at least judged my approach to be unduly transcendental. Having to distill all the conceptual background into reasonably concise and intelligible prose was a tremendous learning experience, as well as a trial of my capacity for self-discipline.

I sincerely hope that several years from now, we will have made sufficient progress in articulating the process of development in natural systems so that this work will then appear rather primitive by comparison. For now, however, I think that this is my best effort at describing whole community development, and it should suffice to initiate dialogue with many who hold dissimilar opinions on the meanings of growth, development, and life.

Robert E. Ulanowicz

Acknowledgments

I am indebted to a number of friends and associates for their help in formulating this work. Perhaps the individual who most directly facilitated this thesis is my close friend and colleague, Trevor Platt. To my memory, Trevor was the only individual ever to respond to a precursory statement of these ideas, which I included at the end of a 1972 publication. Several years later, when I was struggling with numerous professional distractions, Trevor managed my appointment to SCOR Working Group No. 59 and encouraged me to use the new position both as justification to devote more time to the study of thermodynamics and development and as an international platform from which to make known my findings. It was while doing background research for the Working Group that my notions on growth and development took on their present, concrete mathematical form. Chairman Kenneth Mann and other members of the Working Group provided me welcome encouragement and criticism during those crucial early months.

Whereas growth may be quick, maturation is often a slow and agonizing process. During the course of writing this book, my chief source of counsel has been my very good friend and collaborator, Alan Goldman. Seldom have I encountered more penetrating insights, creative alternatives, or incisive criticisms than those which Alan has provided me. Citations of our common work are scarce, but that certainly should not be a gauge of our significant cooperation, from which I have greatly learned and richly benefitted.

Particular mention is also due several individuals at The Institute of Ecology at the University of Georgia. Bernard Patten was especially helpful to me during the revision of my first draft—not only through his own painstaking review of the work, but by the use of this text in his course on advanced systems ecology. Thomas Burns, Lee Graham, Masahiko Higashi, and Thomas James, students in that course, were kind enough to supply me with written critiques. I owe a special salute to a man whom I have never had the honor of meeting, Eugene Odum. It was reading his *Fundamentals of Ecology* which moved me to change careers and pursue the study of ecology. In his only missive to me, he was exceedingly generous in his assessment of my attempts to synthesize his observations on ecosystem development, an act that gave me the needed confidence to continue on what, to many, seemed a radical course of thought.

Over the last few years, Simon Levin of Cornell University has guided several of my journal articles into print, and he was a most welcome catalyst to the publication of this work by Springer-Verlag.

A number of other colleagues read and commented on parts (and in many cases all) of the draft manuscript. Henri Atlan, Albert Cheung, Michael Conrad, Richard Emery, James Kay, Bruce Hannon, Robert May, Ian Morris, Robert O'Neill, Eric Schneider, Janusz Szyrmer, and Richard Wiegert gave most generously of their time and advice. Beyond reviewing my rough draft, these individuals have done much through our personal associations to influence my career and thinking. I should also add that I have not always followed the advice they offered, so that any shortcomings of this volume should be ascribed entirely to me.

My financial support while writing this text has come primarily from my own organization, the Chesapeake Biological Laboratory. I have been most fortunate to work for a series of administrators and with a host of colleagues who, for the most part, have been very indulgent about my pursuit of development theory at times when it was decidedly not in their own fiscal interests to do so. The Laboratory also provided me with the able secretarial services of Gail Canaday, who used her excellent skill on the word processor to produce a text that amounted to typewriter artistry. Frances Younger executed the illustrations with both proficiency and enthusiasm. Numerous ideas in the last two chapters were developed while I was partially supported by grant No. ECS-8110035 from the Systems Theory and Operations Research Program of the National Science Foundation.

On a personal note, the roots for this theory go back over a quarter century while I was still in school and living at home. My parents, Edward and Mary, provided me with an environment extraordinarily free from those concerns (and often responsibilities) that might have otherwise interfered with the freedom to indulge my every intellectual fancy. May they find in this work some small token of my appreciation for their labors and love.

Lastly, and most certainly not least, it is traditional for an author to thank spouse and family for their patience and support. These I have received in abundance. But my wife, Marijka, was far more than just supportive. She labored diligently with me in reworking the first draft—polishing the syntax, testing my explanation of ideas. Her influence is found on almost every page, and it is with affection and gratitude that I dedicate this work to her.

Contents

1
Introduction

". . . induction, perhaps, is not demonstration any more than is division, yet it does make evident some truth."

Aristotle
Analytica Posteriora

1.1 The Enigma

Living things grow and develop. How fundamental is that observation to human experience! Well before the child knows what planets or atoms are, he is made aware of changes within himself and in the living things that surround him. Growth and development, so readily observable, are among the first concepts to establish themselves in the juvenile mind.

Yet how awkwardly this reality sits with what the individual comes to know as science. Science is at its best when it affords an explanation of what has been observed and a prediction of what might happen. Strictly speaking, an explanation can be made in terms of events at scales either smaller or larger than the phenomenon taking place. But in biology, where growth and development are most frequently encountered, there has been a decided bias towards explanations in terms of the smaller, or microscopic, realm. Bubonic plague is traced to rats, whence to fleas, and finally to a causal bacterium. Eye color is traced to a certain pattern of DNA molecules in the genetic material of the parents. One measure of the success of such discoveries has been their utility in predicting and controlling the observable world, often to the great benefit of humanity.

Explanations that have attempted to assign causality at larger scales have not fared nearly as well. Vitalism, teleology, and Lamarckism, to name three well-known historical arguments for nonreductionistic causality, are commonly considered to have been discredited. So great, in fact, is the disdain for these early attempts at biological explanation that considering nonreductionistic causality still appears taboo to the majority of biologists.

This self-imposed stricture to emphasize only influences from the microscopic world implies that growth and development also have their ultimate origins in the molecular realm. Such perspective was significantly strengthened with the brilliant elucidation of the chemical and morphological structures of DNA. The results of specific molecular changes potentially could be mapped at the level of the organism. Molecular determinism achieved even greater stature.

In spite of these striking accomplishments, growth and development remain enigmatic. The mature living entity is generally a system of enormous complexity in comparison with which the DNA molecule, or even the entire germ cell, appears appallingly simple. How can the information to determine such a diverse structure possibly be encoded in the simple matrix of the genome?

The enigma only deepens when one considers that growth and development are not confined to ontogeny. Ecosystems, economic communities, social structures, cultural movements, and even meteorological and galactic forms are perceived to undergo what is commonly regarded as growth and development. Some biologists are still attempting to explain the evolution of large-scale systems in molecular terms. Thus, the proponents of sociobiology (see Wilson, 1975) view genetic factors as controlling the behavior of social communities. Such an attitude seems peculiar to the study of biology. Most meteorologists would consider it a waste of time to attempt to explain the formation of hurricanes solely in terms of the properties of water molecules.

1.2 The Imprecise Universe

Of course, no one doubts that genes do influence social structures, or that the properties of water are very important in the evolution of hurricanes. What is at issue, however, is the magnitude of the effect that any single causal factor may have in the realm of natural phenomena. Just how unambiguously can the consequences of a particular molecular event be manifested at higher scales?

Here an historical parallel is relevant. Intellectuals of the late 18th Century were enamored of the phenomenal success of newtonian mechanics. Perhaps the highest accolade a natural philosopher could receive was to be called a mechanic. The laws of mechanics were considered to be confined not only to machinery and moving bodies, but also were seen at work in economics, social contracts, and governments as well (Wills, 1978). About the same time, the atomic hypothesis was receiving greater acceptance because of its utility in the theory of chemical transformations. The evolving picture of the world (especially gases) was one made up of rigid atomic particles moving about according to the laws of Newton.

To be sure, Newton's laws are quite precise. They give a clear deterministic picture of what happens under prescribed conditions. The speculation, therefore, arose (Laplace, 1814) that if a supernatural "demon" could know the precise location and momentum of every atom in the universe, then Newton's laws would reveal the entire course of cosmic events, past and future.

Laplace's demon seems humorously naive in the hindsight of any 20th Century physicist, but during the enlightenment, it was a very seductive notion reflecting the limitless optimism everyone held for mechanics. Evidence that something was gravely amiss with Laplace's demon began to appear in the 1820s. Engineers such as Sadi Carnot were experimenting with the new steam engines and exploring how efficiently they could be run with a given quality of steam. They discovered empirically that some of the energy in the steam was inaccessible. This implied that the whole process of running the steam engine could not be reversed.

Irreversibility seems so self-evident in retrospect—as inevitable as death and taxes. However, its formal description in macroscopic terms was to become a death blow to Laplace's demon. Newton's laws, after all, are reversible with respect to time. A moving picture of the motions of newtonian bodies looks the same if run either forward or backward; one would not know which direction was forward, nor which backward. How could an aggregate of such reversible processes become irreversible?

Eventual resolution of this dilemma came through the discovery of the laws of quantum mechanics. In short, Newton's laws were shown to be incomplete descriptions of events at the molecular level. Atoms, it seemed, behave as much like waves as they do billiard balls. An implication of this dual nature of material was Heisenberg's principle that it was impossible to know both the location and momentum of a particle to more than a finite level of precision. Not only is the universe irreversible; it can never be precisely known.

Imprecision tends to grow in any large-dimensional system—especially if the dynamics are nonlinear. There are innumerable examples of systems of equations, such as those describing the many-body problem, that appear to be deterministic; but in reality, they give rise to behavior that cannot be distinguished from chaos (Lorenz, 1963; Ulanowicz, 1979). That is, if one could know with infinite precision the starting conditions for the system, it would always follow a predicted set of trajectories. However, any deviation, be it ever so infinitesimal, will in time take the system to a configuration that bears little resemblance to the predicted state. If only one particle among the countless number making up the universe were not known with absolute precision, the demon's prediction (control) would eventually go awry. Conversely, it becomes impossible to retrieve states far enough in the past; that is, the system cannot be precisely reversed.

1.3 The Dilemma of Modern Biology

At first glance, it would seem that recent biology bears little relation to Laplace's demon. After all, chance plays a large role in the theory of heredity. But outward appearances can be deceptive. There was an obvious trend in the thinking of biologists during the 1950s and 1960s towards exhuming Laplace's creation.

The rewards and the taboos that encouraged reductionism have already been mentioned. The culmination of this tendency was the attitude (still held by many today) that causality can flow only *up* the hierarchy of entities and events from smaller to larger scales. Chance may act directly at the molecular level, but from there, cause was projected in a precise fashion towards higher levels. Although random events could occur at higher levels, their significance could only be understood if one related them back to the molecular world. For if any aspect of macroscopic structure were in some way autonomous of its molecular antecedents, then its organization would be the result of some law or agent operating at a higher level. Orthodoxy was therefore maintained by telescoping DNA as the shaper of social structures, or by regarding the organism as a mere mechanism for perpetuating its particular form of DNA.

There is now growing recognition by molecular biologists themselves that the arguments for "genetic determinism" may have been overdrawn. The most convincing grounds for such disillusionment have been the difficulties encountered in recent empirical attempts to "map out" the exact pathways over which the genome is expressed in the phenome.

Lewin (1984) reported on the efforts of Sidney Brenner and associates to specify completely all genetic influence in a very elementary multicellular organism, *Caenorhabditis elegans*, which is a tiny nematode. With only 959 cells in the body of this creature, it is conceivable to catalogue the precise effects of known genetic substitutions. However, after 2 decades of elegant planning and excellent execution, "an understanding of how the information encoded in the genes relates to the means by which cells assemble themselves into an organism . . . still remains elusive."

In Brenner's own words: "At the beginning it was said that the answer to the understanding of development was going to come from a knowledge of the molecular mechanisms of gene control . . . [But] the molecular mechanisms look boringly simple, and they do not tell us what we want to know. We have to try to discover the principles of organization, how lots of things are put together in the same place."

It is not as though genetic determinism were being pushed aside by an ascendent theory of organization. Despite the existence of several intriguing theories of development (Prigogine and Stengers, 1984), those biologists who are aware of the importance of macroscopic development have not yet settled upon a suitable theory. This places the typical biologist in an acute dilemma. One can choose (as, indeed, many do) to adhere conafield from biology into other realms where growth and development might be considered to occur (e.g., into the fields of economics, sociology, or cosmology mentioned earlier). As chance would have it, ascendency is an outgrowth of concepts and observations now popular in eco-

minefield planted with vitalism and teleology. It is little wonder that so many would like to avoid the issue altogether!

1.4 Phenomenological Redress

The cause of this apparent dilemma and the key to its eventual resolution can be traced to a perceived imbalance in the quality of descriptions of events as they occur at the micro and macro scales. The narrative of molecular biophysics is certainly fascinating and has been woven tightly into what is commonly referred to as the neo-Darwinian hypothesis. By contrast, the collection of observations at the macro scale often appears disjointed. There is virtually no agreement on a core principle to which the various macro phenomena can be referred. It is as much by default, as by any causal ties, that higher level phenomena are still usually referenced back to biomolecular events.

This book is an attempt to begin to remedy the present imbalance in theoretical description. Just as the ardor for Laplace's demon was cooled by the birth of thermodynamics, resolution of the present dilemma is sought here in terms of improved phenomenological description. Phenomenology is the description of the formal structure of the objects of awareness in abstraction from any claims concerning existence. In this text, growth and development—those features so fundamental to living systems—are characterized by a new, *quantitative* formalism of increasing "ascendency." The observable drives of living systems towards coherency, efficiency, specialization and self-containment are argued to be implicit in the "principle" of optimal ascendency. The ultimate aim is to provide a synthesis of concrete observation to which nonreductionistic ideas can be referred.

However, where and how does one begin the exposition of a coherent formalism for growth and development? The noted developmental biologist Gunther Stent (1981) offers a lead: "Consider the establishment of ecological communities upon colonization of islands or the growth of secondary forests. Both of these examples are regular phenomena in the sense that a more or less predictable ecological structure arises via a stereotypic pattern of intermediate steps, in which the relative abundances of various types of flora and fauna follow a well-defined sequence. The regularity of these phenomena is obviously not the consequence of an ecological program encoded in the genome of the participating taxa."

Stent is suggesting that it is necessary to distance oneself from the bimolecular and organismal realms, but not to proceed at first too far servatively to the strict, reductionist view of nature and to continue paying homage to Laplace's demon. Or one can proceed boldly to still talk in nebulous terms about principles of organization that extend beyond molecular mechanisms, all the while threading deftly through a conceptual

systems science, so it is only appropriate that this description of growth and development proceed in the context of ecology.

It is hoped this text will also contribute an apologia for ecosystems science. After all, if causality were to issue only from molecular agents, why devote one's career to the pursuit of corollary knowledge? Rather, most investigators of ecosystems are convinced that what they are describing has significance in its own right. As Stent suggests, it is not necessary, and very likely impossible, to interpret ecosystems phenomena solely in biochemical terms. Over the long run, in fact, ecosystems can be perceived to develop as a unit and to influence strongly their own biochemical structure.

1.5 Origins of the Principle

Like the evolution of systems described in this text, new principles do not appear *ex nihilo*. Even revolutionary advances have always borne some relationship to the body of scientific knowledge that preceded them. In fact, the acceptability of new ideas is often judged by how much more coherent the corpus of science becomes after the new concept has been added. It is necessary, therefore, to first study the elements that go into the eventual description of growth and development.

To begin with, one must decide upon a *perspective;* that is the rules that govern the game to be played. In this regard, the phenomenological perspective of thermodynamics is adopted. Most readers will probably have some familiarity with thermodynamics, but a perusal of introductory texts on the subject reveals little discussion of the underlying phenomenological foundation. The second chapter is an attempt to provide some treatment, however cursory, of these vital preconditions.

Having decided how to look, the question next arises as to what the *object* of study is to be. Granted, growth and development are the ultimate goals for description; but these phenomena appear in various guises when treated by different disciplines. Does a common facet of the diverse forms of growth and development exist, and can that aspect be adequately measured? In Chapter 3, networks of flows are proposed as the most promising objects to be quantified.

Anticipating that the universal features of ecosystem network development can be catalogued, it is only natural to ask the question: "What causes lie behind the description?" In Chapter 2, it will discuss how answering this question is really outside the realm of phenomenological science and, thus, the scope of this text. However, unless the reader is supplied with at least one class of *agents* or mechanisms, his inclination will probably be to believe the author has only supernatural constructs in mind. Therefore, to prevent this exposition from taking on an excessively metaphysical flavor, cybernetic feedback, as it is associated with material and energy cycling, is proposed in Chapter 4 as an example of a nonreductionistic agent that can effect growth and development.

It remains only to transform the measurements on developing networks into a set of mathematical indices that sufficiently characterize growth and development while implicitly addressing the agents behind them. For reasons that later become clear, information theory is introduced in Chapter 5 as a *calculus* for the appropriate manipulation of measured flows.

The stage is finally set for the attempt to characterize growth and development. Four chapters of preliminaries are a lot to ask of any reader, but such preparation is necessary if the thesis is to be fully appreciated. The introductory materials have been abbreviated as much as possible, and those wishing in-depth introductions to the four subject areas are forewarned that they will not be found here. Rather, each introductory chapter is intended to give those unfamiliar with the subject a heuristic feel for the major elements of that discipline and to emphasize subtle points (often not mentioned in introductory textbooks), which are important to understanding this synthesis. Hence, even those who feel wellgrounded in one or more of the four subject areas are likely to find new aspects of familiar topics.

To grasp any quantitative theory requires some mathematical background on the part of the reader. However, every conceivable effort has been made to minimize the amount of mathematics necessary to follow the text. At this point, the reader is expected to know only algebraic manipulation, the rules of matrix algebra, and the fundamental properties of logarithms (e.g., $\log AB = \log A + \log B$). Some of the manipulations in the later chapters are tedious, but it is hoped that by that time the reader will be sufficiently motivated to invest the time to check those derivations. In attempting to avoid the introduction of differential calculus, the next chapter on thermodynamics might appear emasculated; but it is the conceptual underpinnings of thermodynamics, rather than the detailed results, that are crucial to the development of principles that follow.

Once the background material has been elaborated, the quantification of growth and development as a unitary process follows in the beginning sections of Chapter 6. The remainder of the book is devoted to demonstrating the power of this definition so laboriously crafted. As an immediate corollary to the definition of ascendency, the same formalism permits the quantification of countervailing processes, such as those towards diversity, variability, senescence, and dissipation—factors that serve to limit the progression of growth and development. The ability that optimal ascendency gives one to unify seemingly unconnected (and sometimes disparate) observations and hypotheses is portrayed; the potential of the theory to serve as a tool in formulating other precise system-level definitions is then demonstrated. One such new concept provides a possible extension to Darwinian thought. "Fitness" as used by Darwin has always prompted the question: "Fit for what?" Community ascendency imparts an appropriate direction to the fitness of a population without necessarily implying a fixed goal in the teleological sense.

The advantage in formulating a definition that follows naturally from existing theories is that extensions and generalizations to the new idea are thereby greatly facilitated. Amending the notion of ascendency in Chapter 7 to incorporate spatial heterogeneities, temporal changes, or material differences presents no overwhelming conceptual difficulties. The final effort to show how the thesis might be applicable to nonecological systems becomes an appropriate anticlimax. The reader who has remained with the development to this point should need little convincing that the scope of the principle is very broad and should be well on the way towards making his own extrapolations.

The view of growth and development described here has substantially heightened the sensitivity of the author to integrated structures. To this writer, the world now appears considerably less disjointed. Possibly, this is only an illusion, but the vision is so compelling as to make it impossible not to want to share it with as many as are willing to take the time to read, think, and react.

2
The Perspective

> "In the dialogue between experiment and theory there are these useful middlemen, the phenomenologists, who try to encapsulate experimental data in suggestive formulae, without being able to give a fundamental explanation for the form chosen."
>
> John Polkinghorne
> *The Way The World Is*

2.1 Thermodynamics: The Phenomenological Science

If there is one chapter in this book about which the reader is likely to feel uneasy, it is this one. Deep down, most scientists feel insecure about their background in thermodynamics. (And most of those who do not probably should!) But from where does this anxiety issue? Certainly, the fundamental laws of thermodynamics are not that difficult to comprehend.

A second discomfort often associated with the subject is disappointment. Students engrossed in thermodynamics are often quite enthusiastic about the insights gained. Some even wax ecstatic over the cosmological implications of the principles they are learning. Most rightly feel thermodynamics to be an invaluable addition to their background, and they look forward to the time when they can apply this knowledge in their problem-solving careers. That time, however, is usually slow in coming. In comparison with such subjects as for example, biochemistry, transport phenomena, or optimal control theory, thermodynamics does not appear as an often-used "tool" in their repertoire. Some have gone as far as to characterize thermodynamics as the laws of impotence (Conrad, 1983). Disappointment ensues. And yet without thermodynamics, these other "tool subjects" are quite incomplete.

The root of both of these subjective difficulties with thermodynamics appears to lie in an incomplete appreciation for the phenomenological nature of the discipline. Simply put, thermodynamics is a codification of what has been experienced (in a quantitative, objective fashion) by a broad spectrum of scientific investigators. The very same juxtaposition of phenomena observed in the 19th Century by engineers working with steam engines is seen to occur without fail in diverse modern areas of study such as mechanics, electronics, chemistry, physics, biochemistry, and ecology. The identical principles can be realized by any one of a limitless class of objective experiences. It is no surprise, then, that a course in thermodynamics designed for students in electrical engineering would be cast in terms of phenomena quite different from those used by someone teaching the subject to molecular biologists, for example.

The variations in thermodynamic expression can extend beyond the experiential base to include the quantitative language (i.e., mathematics) used in the descriptions. The classical approach is cast mostly in terms of partial differential relationships, whereas the highly abstract exposition of Caratheodory (1909) draws heavily upon rational analysis (in the mathematical sense) and topology. The approach of Tribus (1961) relies upon statistics and information theory.

At present, it is unrealistic to expect anyone to be very familiar with the phenomena peculiar to every discipline. Likewise, it is rare to find an investigator highly proficient with all pertinent mathematical tools. The result is that one is constantly encountering thermodynamic arguments cast in highly unfamiliar terms—like hearing a familiar verse read in an unknown foreign language. Unease is a frequent result.

As to disappointment, it can be traced to the fact that most introductory texts do not adequately stress the limitations inherent in the phenomenological nature of thermodynamics. As a codification of objective experience, it is highly inductive in nature and does not "explain" matters in the deductive sense of the word. For example, using postulates derived from evolutionary theory, genetics, and molecular biology, coupled with a knowledge about the specific environment in which a species lives, the neo-Darwinist is able to "explain" the appearance of a particular trait or behavior, such as white fur on polar bears or the diurnal migrations of zooplankton in the water column. Thermodynamics offers no explanations and only the most diffuse predictions (e.g., the entropy of the universe will increase). Not having been forewarned, students often build ill-founded expectations for thermodynamics born of their experience with other, more deductive and predictive branches of science. Disillusionment all too frequently ensues.

In light of the various difficulties individuals have with thermodynamics, it is only appropriate to ask what constitutes the underlying power of this discipline. Why is thermodynamics considered to be a cornerstone of science? The answer lies in its exceptional generality or universality. Above a certain scale, the laws of thermodynamics have been observed to hold without exception. Hence, it matters not whether one is dealing with a plasma, an electromagnet, a packed-bed reactor, a soft clam, or a pond ecosystem, the thermodynamic laws remain inviolate. This is a highly significant attribute—no other branch of science can lay claim to such generality. Finding a single system that violated the thermodynamic laws would be grounds for rejecting or at least significantly rewriting them. It would be a discovery of the first magnitude.

Universal description is possible only if one views the world in macroscopic fashion. That is, one assumes the systems being observed consist only of one or a few elements and deliberately ignores the internal, or microscopic, structure of these elements in any detail. Thus, if one is concerned with a gas in a cylinder, it is sufficient to know the gross properties of the gas—its temperature, volume, and pressure, for instance. Certain situations may require a knowledge of other properties, such as viscosity or thermal capacity; but under no circumstances is it imperative to know that the gas is made up of atoms and molecules, much less to know exactly what those molecules might be doing.

Many individuals remain suspicious of the independence of thermodynamic principles from a knowledge of microscopic mechanisms. This confusion arises from a misassessment of the role of statistical mechanics in science. In statistical mechanics, one starts with a set of assumptions concerning the properties of "atoms" that constitute a particular material. For example, the molecules of a gas may be assumed to be point masses, may be postulated to behave as hard spheres with a given radius, or may be hard spheres with an attraction potential for one another that falls off as the square of the distance from the center of the molecule, etc. The atoms are then considered to be distributed in location and momentum according to prescribed or empirically derived statistical distributions. Invoking the laws of mechanics and statistics to calculate the aggregate behavior of the ensemble yields approximations to such empirical "laws" as those for a perfect gas or a Van Der Waals gas (Chapman and Cowling, 1961).

To a very limited extent, statistical mechanics provides an "explanation" for thermodynamic phenomena under a very narrow range of conditions. This leads to a feeling in some that thermodynamics may be deduced from more fundamental elements. To think in this way is to invert priorities. Thermodynamics is a self-consistent body of empirical knowledge that may be readily verified without any recourse to microscopic particles. The aim of statistical mechanics was to lend credence to the atomic hypothesis by demonstrating that it could be related to the *more fundamental* empirical laws. A failure of this effort would have meant trouble for the microscopic hypothesis, not for thermodynamics.

What statistical mechanics does accomplish (aside from deriving some handy formulae for estimating the properties of matter) is to establish the mutual compatibility of thermodynamics and the atomic hypothesis, and therein lie several helpful insights.

In proceeding from the behavior of the very many elements at the microscopic level to the actions of the few properties of the macroscopic, the many-to-one correspondence is made explicit. Under such homomorphic relations, a single given value of a macroscopic property could be a manifestation of any numerous, different configurations at the microscopic scale. Were it possible to see the molecules in a fluid, the vision would be quite chaotic; not only in the subjective sense, but in one's inability to predict the evolution of the configurations far into the future (Ulanowicz, 1979). Yet, macroscopically, the fluid may be behaving in a very regular and predictable manner. The appearance of chaos and degrees of freedom at the individual level do not preclude the orderly and deterministic behavior of the ensemble.

With these remarks as background, it is not difficult to see how the application of thermodynamics to ecosystems analysis and social systems might be fraught with difficulties. To begin with, it is unclear whether the body of quantitative observation at the macrosystem level is yet sufficient to allow for a thermodynamic description in ecological terms. Similarly, the language (mathematics) used in physical thermodynamics may not be the most appropriate for describing community phenomena. As will be argued, most of what now passes for ecosystems thermodynamics consists of attempting to use classical state variables (derived for equilibrium situations) to describe physiological processes at the scale of the organism.

Now, when one looks upon an ecological community, the "atoms" of the system appear directly as tangible objects. The armadillos, the trees, the euphausiids, the beetles, and the zooplankton move about in irregular patterns and engage in any number of discrete interactions at apparently random intervals. That is to say, humans are born into the microscopic world of ecology replete with chaos as well as splendor. The crucial challenge is to obtain or infer a macroscopic vision of the ecosystem (Margalef, 1968), and the key question is whether such vision will appear orderly and amenable to description.

An early effort at such macroscopic description was made by Kerner (1957), who attempted to create a strict analogy between ecology and statistical mechanics (see also Goel et al., 1971). Although there has been no serious attempt to measure Kerner's global variables, there remains skepticism as to how useful they might be, given the stringent assumptions necessary to define and compute them. (For example, only predator-prey type interactions are allowed.)

If, then, the construct of statistical mechanics seems inadequate to the task of formulating systems-level properties, what are the alternatives? One novel approach is suggested in the succeeding three chapters. Pertinent here are the restrictions that the foregoing discussion places upon this subsequent development. For example, the power of the description must lie in its universality. That is, the thermodynamic description of organization in ecosystems must have the same form as the description of organization in for example, ontogenetic or economic systems. Prohibited, of course, is any descriptor that relies upon specific mechanisms. Hence, the mechanics of DNA replication could not appear explicitly in a macroscopic description of ecosystem organization. (Which is not a value judgment upon molecular biology, the results of which are almost certainly compatible with macroscopic descriptions in the same sense that classical thermodynamics and the atomistic hypothesis are reconciled by statistical mechanics.)

Although the phenomenological perspective offers great hope for a non-reductionistic approach to ecology and other large-scale complex systems, it must be acknowledged that the present body of thermodynamic theory is still insufficient to provide an adequate description of the phenomena of growth and development as they exist in communities (Lurie and Wagensberg, 1979; Johnson, 1981; Gladyshev, 1982 for opposing opinions). A successful description would then constitute an addition to the framework of thermodynamics. Certainly, this additional principle cannot reverse the existing laws of thermodynamics, but rather must be compatible with them. It is necessary, therefore, to first discuss the accepted body of thermodynamic definitions and concepts necessary to provide a background for the theory to follow.

2.2 The First Law and the Nature of Work

Practically everyone with any scientific training has been exposed to the principle of the conservation of energy. Energy can neither be created nor destroyed, only changed from one form to another. It is one of the few principles to endure without exception over all observable scales of space and time.

The principle actually marries two concepts—conservation and transformation. Of the two ideas, conservation, was the first to be formulated. The way for thermodynamics was paved in the late 17th and early 18th Centuries by the development of the thermometer, allowing observers to assign a given amount of *caloric* to a body in proportion to both its mass and its temperature. The evolution of the concepts of conservation of both material and chemical elements was paralleled by the development of the conservation of caloric (Tisza, 1966). Thus, if a mass of water m_1 heated to a temperature θ_1 were mixed with a mass m_2 at temperature θ_2, the caloric was assumed to be conserved, that is,

$$m_1\theta_1 + m_2\theta_2 = (m_1 + m_2)\theta \qquad (2.1a)$$

where the final temperature of the combined volume θ is calculated as

$$\theta = (m_1\theta_1 + m_2\theta_2)/(m_1 + m_2). \qquad (2.1b)$$

This inchoate notion of constancy of caloric was refined to allow for the specific heats of differing substances, as well as for heats of reaction and latent heats of phase transition; and the caloric theory acquired wide acceptance by virtue of its simple elegance.

So attractive was the caloric theory that most adherents did not abandon it in the face of Count Rumford's turn-of-the-century demonstration that caloric could be created from mechanical work (boiling water by means of the frictional heat created while boring a cannon). It was another 40 years or so before R.J. Mayer and J.P. Joule independently estimated the equivalence factor for the conversion of work into heat. In fact, through exacting experiments, Joule demonstrated the interconversion of mechanical, electrical, chemical, and caloric energy. The concept of the indestructible caloric began to lose favor, but the calculus of conservation was retained. To calculate the change in energy of a system (ΔE), one had to account for both the flow of caloric into the system (Q, heat) and the work done on the system by the surrounding universe (W) in the manner,

$$\Delta E = Q + W \tag{2.2}$$

One may well ask whether the discovery of energy transformation almost 150 years ago has had a truly significant impact upon the way ecological energy budgets are presently reckoned. Usually, the unit of the system under consideration (a species, a population, a trophic level, etc.) is depicted like the box in Figure 2.1. The inputs to the compartment from various sources are shown by the arrows entering from the left and labelled T_{ji} ($j = 1, 2, 3, \ldots, n$). Outputs associated with biomass flows out of the component are labelled T_{ij} ($j = 1, 2, 3, \ldots, m$) and exit to the right. Sensible heat leaving the compartment is represented at the bottom of the box by the arrow labelled Q_i.

Most ecological energy balances proceed along the following lines. First, the biomass flow rates entering and leaving the compartment are estimated by any one of a number of methods. Then the energy per unit biomass is determined using bomb calorimetry. This procedure consists of drying a specimen, placing it inside a strong-walled metallic vessel ("bomb") inside a water bath, injecting oxygen into the bomb, igniting the contents of the bomb, and measuring the consequent rise in temperature of the water bath. Knowing the mass of the specimen and the mass of the water in the bath, one may then calculate the heat of combustion of the specimen per unit weight. (The quantity measured by this method is actually the enthalpy per unit mass, but the distinction is not crucial to what follows.) Each biomass flow is then multiplied by the corresponding energy per unit biomass to obtain the energy flows entering and leaving compartment i. The heat flow Q_i (respiration) is estimated either by some

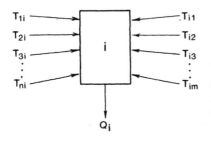

FIGURE 2.1. Typical inputs and outputs of energy to an arbitrary compartment i. Over sufficient time, inputs must always exceed outputs by an amount Q_i, which is dissipated as heat.

sort of calorimetry experiment, wherein the evolution of heat from the animals is directly sensed, or by measuring the rate of oxygen uptake by the organisms and converting that rate into heat via the known catabolic heat of reaction.

As energy is neither created nor destroyed by any of the processes internal to compartment i, one may write the instantaneous energy balance,

$$\begin{pmatrix} \text{rate of accumulation} \\ \text{of energy by } i \end{pmatrix} = \begin{pmatrix} \text{sum of all energy} \\ \text{fluxes into } i \end{pmatrix} - \begin{pmatrix} \text{sum of all energy} \\ \text{fluxes out of } i \end{pmatrix}$$

or,

$$dE_i/dt = \sum_{j=1}^{n} T_{ji} - \sum_{k=1}^{m} T_{ik} - Q_i \qquad (2.3)$$

where E_i represents the energy internal to i, and the prefix d/dt indicates the time rate of change. Most biological processes occur under conditions of nearly constant pressure, practically assuring that the mechanical work effected by compartment i is negligible and would not appreciably affect the balance in (2.3).

The reason for belaboring the process of energy balance in ecosystems is to point out that there is hardly anything in the entire procedure that does not fall under the aegis of the caloric theory. That is, an experimenter ignorant of the interconversion of heat and work (and the differences between internal energy and enthalpy), in most cases, could go through the same measurements and arrive at an energy balance that would not significantly differ from (2.3). Might Mayer and Joule just as well have spared their labors? Is the concept of work of such little significance to ecological analysis?

Of course, the notion of work is pertinent to ecosystems; but to see why this is true, it is necessary to briefly regard aspects of work other than just its dimensionality. In order to initially formulate the first law, it was necessary to focus upon the physical dimensions of work. The many different forms of work (e.g., a mechanical force acting through a distance, an electrical charge moving through an electrostatic field, and a number of moles of substance transferred to a different chemical potential) can all be measured in the same physical dimensions, which are identical to those of heat. The equivalency of the different forms of work is a major discovery built into the first law, and there the matter is usually left to lie. However, as will be seen in the next section, interest in heat extends beyond its dimensions, and much attention is paid to the quality of thermal energy. To lapse for a while into the microscopic, the qualitative aspect of heat in statistical mechanics may be thought of as the transfer of thermal energy from a medium with a molecular configuration (in both space and momentum) that has a probability of occurrence small in comparison to the more probable configuration of the receiving medium. Transition proceeds from the less probable to the more probable.

In trying to imagine work on a microscopic level, it becomes obvious that its qualitative nature is opposite to that of heat. To raise a 1 kg stone from the bottom of a hill 100 meters high to the top may result in 981 joules of work performed, but it is also true that the probability of the stone's being at the top of the hill is smaller than the probability of its being at the bottom. The removal of all constraints keeping the stone at the top will result in the transition of the stone to the bottom. In the absence of constraints, spontaneous events will occur in the direction of the higher probability (a disguised statement of the second law). Imagining any performance of work in reverse will depict a spontaneous process. The compression of a gas in a cylinder requires work. Let go of the piston and the gas spontaneously expands to a larger volume. Conversely, it requires work to proceed from a state of higher probability to one of lesser likelihood. If two dissimilar gases are separated by a partition and the partition is removed, the gases will become a single uniform mixture. To reorder the gases into their separate compartments requires a calculable minimum amount of work (chemical work, to be more specific). It is in such examples that work is seen to be an ordering process. In this sense, the scientific meaning of work is broadened to agree more with the everyday notion. The laborer stacking bags of concrete and the clerk filing papers are both engaged in thermodynamic work. The only factor differentiating the two tasks is the degree of difficulty it takes to quantify the magnitudes of their respective work in units of energy.

Returning to the compartment in Figure 2.1, it can be argued that work is being done. Again lapsing into the microscopic world, the molecular configurations of the input flows are being rearranged into new forms as they appear in the output flows. The chemical work associated with such rearrangement could be calculated if it were possible to measure the chemical potentials of the inputs and outputs and the number of moles of organic matter undergoing transformation. However, measuring the chemical potentials of simple substances is difficult enough, let alone measuring the incredible mixture of organic substances in any ecological flow (Scott, 1965). Besides, the chemical (or biochemical) work is only a part of the total work process. To determine the chemical potentials of the constituent organic compounds, it becomes necessary to separate them from the matrix of the living organism. In turn, the organism must be extracted from the network of exchanges with other individuals and species. One expects, then, that macroscopic biological and ecological work functions exist associated with the ordering inherent in these living networks. Perhaps surprisingly, it may be easier to estimate the larger scale (ecological) work function than to measure the work performed at either the chemical or organismic levels, but a description of such estimation awaits the development of concepts introduced in Chapter 6.

2.3 The Second Law

If the discussion about the nature of work required that one peek ahead to the second law, then such inconsistency is perhaps forgivable in light of the fact that the second law of thermodynamics was articulated some 20 years prior to the first law by the French military engineer Sadi Carnot. Carnot (1824) was concerned with the design and operating characteristics of early steam engines used at the time principally for pumping water from mines. He actually began his analysis with the statement of the second law in the form: "It is impossible to construct a device that does nothing except cool one body at a low temperature and heat another at a high temperature" (Tribus, 1961). He then turned his attention to the various cycles of heating, decompression, cooling, and compression to which the steam in the engine could be subjected. By using his version of the second law, he was able to show that there is a particular cycle that is more efficient (in terms of converting heat to work) than any other imaginable. To this day, the efficiencies of working machines are still compared to the Carnot efficiency.

There are multitudinous later statements of the second law that can be shown to be equivalent to the Carnot postulate. Clausius, for example, restated Carnot's notion as: "Heat cannot of itself flow from a colder body to a warmer one." This is a statement that might sound like folk wisdom to today's reader, but one must remember that during the middle of the 19th Century, the nature of temperature as a potential for heat flow was just being realized. However, some statements of the second law became rather abstract. Witness Caratheodory (1909), who in the language of differential geometry interpreted the second law as requiring that arbitrarily close to any point in phase space (the mathematical space defined by the state variables defining an equilibrium system; e.g., temperature, volume, and pressure) there exists an infinite number of points that cannot be reached by a reversible process.

The form of the second law that seems most appropriate to ecosystems analysis is that it is impossible to convert any source of energy entirely into work. Referring back to Figure 2.1, one pictures T_{ji} as being a source of energy for ecosystem component i. In the larger sense, the ecosystem work performed by the compartment is measured by the sum over k of the T_{ik}. The second law requires that the total input exceed the aggregate work by some positive quantity dissipated to the universe as heat, or

$$\sum_{k=1}^{n} T_{ki} - \sum_{k=1}^{m} T_{ik} = Q_i > 0. \qquad (2.4)$$

Every functioning component of the ecosystem will possess an obligatory rate of dissipation Q_i, which will be accorded a special symbolism in future diagrams akin to the ground symbol in electrical diagrams (Figure 2.2).

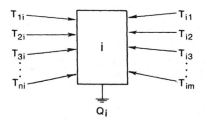

FIGURE 2.2. Typical inputs and outputs of energy as in Figure 2.1. Here the dissipative flow, Q_i, is represented by the ground symbol.

There is some question as to how rigidly the second law constrains mass balances in ecosystems. While the flow Q_i in an energy balance represents the departure of heat no longer available to do work, all forms of a material element have the potential for feeding back into the network. Hence, many investigators draw networks of material transfers without any losses to the external world. Skirting the issue of whether such a representation is proper or not, it will be the convention in the remainder of this book to identify one form of the element under consideration as being the most degenerate state of that medium (e.g., the form having the least available energy). This ground state will not appear as a compartment in the network of exchanges, and flows to the ground state will be represented by the ground flow symbol. Reentry into the network requires a source of energy, and the accompanying mass flows will be treated as primary inputs to the ecosystems. For example, if carbon is the element being circulated, the respiratory flows (ground symbols) represent the degradation of other forms of carbon to CO_2, whereas the fixation of CO_2 into organic matter via photosynthesis will appear as exogenous inputs to the producer compartments from outside the system.

The foregoing discussion of the second law has been sufficient to treat the types of measurements now being made on ecosystems. Again, the thermodynamicists of 130 years ago were already equipped to interpret such data in much the same way as done now. However, even the casual student of thermodynamics would be quick to point out several later ideas pertaining to the second law about which no mention has yet been made.

Several of the concepts key to the development of equilibrium thermodynamics in the late 19th Century are of marginal utility in discussing ecosystem events. For example, one hears of certain macroscopic properties as being essential to a description of the state of a system; that is, they are *state variables*. The classical examples are the pressure, temperature, and volume of a gas. State variables can be either extensive or intensive depending upon whether or not the value of the variable conveys any information about the physical size of the system. Thus, pressure is intensive, in that any imaginary partitioning of the system leaves the pressures of the subsystems unchanged. With volume, however, any partition of the system will apportion the value of this extensive variable to the subsystems. (See Amir, 1983, for another interesting distinction between intensive and extensive variables.)

State variables are often contrasted with process variables, which describe what happens during a transition between states. For example, a system is taken through any imaginary path, so that the starting state and final state are identical (i.e., the path is a cycle). By definition, the temperature (a state variable) is the same in the final state as at the beginning of the process cycle. The heat transferred through the system (a process variable) is not, in general, zero and depends upon the specific pathway of the cycle. It happens that state variables possess certain mathematical properties (they are perfect differentials) that make it convenient to express relationships among them. For this and other reasons, significantly more attention is paid to state variables, and process variables have not received nearly as much discussion in introductory textbooks. This bias results in many students coming away with the feeling that process variables are somehow "flawed" in comparison with state variables.

State variables, however, by their very name imply conditions that are static or unchanging. They strictly exist only under conditions known as thermodynamic equilibrium. Equilibrium in thermodynamics has a much more precise meaning than the general idea of an unchanging balance. Any system with properties unchanging in time is said to be in steady-state. Only if no dissipation is occurring within the system is it also at thermodynamic equilibrium. One can usually differentiate between a system at steady-state and one at equilibrium with the help of a simple thought experiment. In one's mind, an unchanging system is completely isolated; that is, all exchanges, of material and energy with the outside universe are severed. If after isolation the system remains completely unchanged, then it was originally at thermodynamic equilibrium. If *any* changes ensue, then the original system was at steady-state, but not at thermodynamic equilibrium.

Example 2.1

A solid conductor of heat is in contact with an infinite heat reservoir of temperature θ_1 at one end, and it connects with a similar reservoir at temperature θ_2 at the other end. $\theta_1 > \theta_2$, but $\theta_1 - \theta_2$ is small. After a sufficiently long interval, the temperatures at various points along the bar are observed not to change with time, and the temperature profile along the length of the bar is for all practical purposes linear.

The bar represents a system at steady-state. It is not at thermodynamic equilibrium, however, which can be ascertained by imagining that the bar is suddenly encapsulated by a barrier that will allow no passage of either material or energy. One need not solve the equations of heat conduction to guess that the linear temperature profile would thereafter degenerate, so that the bar would eventually reach a temperature of about $1/2(\theta_1 + \theta_2)$ everywhere along its length. Because the system changed after isolation, it was not at equilibrium while in the steady-state.

Example 2.2

A reactor vessel held at a constant temperature of 986°C initially contains a gaseous mixture of 70.3% CO_2 and 29.7% H_2. After sufficient time has passed, the vessel is seen to contain unchanging proportions of 47.5% CO_2, 6.9% H_2, 22.8% CO, and 22.8% H_2O. This final steady-state mixture of gases is at true thermodynamic equilibrium, for if the reaction vessel were immediately isolated, no further change in the mixture would ensue.

Such thought experiments immediately tell the ecologist that the systems he is interested in are never at equilibrium so long as they are alive. How appropriate, then, is it to apply the concept of a state variable to ecological systems? The answer depends upon the nature of the state variable in question. For example, one has little difficulty conceiving of or measuring the temperature of an organism. This is due to the fact that there are usually only slight gradients of temperature within the organism. If one could suddenly isolate the body of the organism in question (and simultaneously shut down its metabolism), the small gradients would disappear and the equilibrium temperature of the isolated system would not differ markedly from measurements made on the living homeotherm or poikilotherm. With respect to temperatures, most organisms (and their aggregates) are said to be "near-equilibrium."

The picture is quite different, however, when one considers applying some other state variables to organisms and communities. Entropy, for example, is the state variable in classical thermodynamics most central to statements of the second law; for example: "In any spontaneous process the entropy of the universe must increase." Entropy in the classical sense is a measure of the unavailability of the energy stored at a given temperature to do work. The higher the entropy of an energy reservoir, the less work can be extracted from it. This notion of availability to do work was captured by Hermann von Helmholz in his definition

$$F = U - \theta S, \qquad (2.5)$$

where U is the total energy of a system beyond that possessed by the system in a reference state, and F is the portion of that energy that can be tapped to perform work. The part of U which is unavailable to do work is quantified by the product of the temperature (θ) and the entropy above the reference state (S).

From the macroscopic viewpoint, the notion of entropy is very abstract and difficult to visualize. However, the microscopic analog, issuing from statistical mechanics, affords more concrete imagery. The mathematical representation of entropy will be left until Chapter 5; suffice it to say here that the entropy of a collection of items comprising a system is a measure of their disorder (or equivalently the relative likelihood of the instantaneous configuration of the items). An often-cited example is that of two different gases separated (ordered) into the halves of a container by a partition. The partition is removed; the gases mix (disorder) into a homogeneous system. The increase in entropy (disorder) for this mixing process can be calculated by either statistical mechanics or by macroscopic methods.

That entropy might provide a quantification of the heretofore subjective notion of disorder has spawned innumerable scientific and philosophical narratives (Kubat and Zeman, 1975). In particular, many biologists have taken to speaking in terms of the entropy of an organism, or about its antonym negentropy, as a measure of the structural order within an organism.

. These efforts at extending classical thermodynamic concepts to living systems have great appeal to the intuition. There are, however, two important reasons to question how legitimate such extensions are. The first concerns the problems of defining a state variable for a system not at equilibrium. Unlike temperature, the putative entropy of a living system would change drastically if the organism were thermodynamically isolated. It would increase markedly as the once-living components decayed to an unrecognizable mass.

Should the pragmatist feel that the first objection borders on semantic quibbling, perhaps he would be more impressed by the second objection that no one has yet been able to measure the entropy of a living system. Measuring the entropy of many physical and chemical entities is no easy task and is complicated by the necessity of assuming a value for the entropy of the substance in some reference state. (Some refer to this necessary assumption as the third "law" of thermodynamics.) As with bomb calorimetry, these methods usually require the destruction of the sample (Scott, 1965). However, as discussed in the last section, the ability of the living system to do work is contingent upon its functioning as a living, dissipative system. Any destructive measurement will perforce neglect these work functions. Hence, many thermodynamic discussions about biological systems are being conducted using constructs that cannot be measured at present. (This includes not only entropy, but derivative variables such as free energy.)

2.4 Nonequilibrium Thermodynamics and Proto-Communities

Classical thermodynamics, in dealing exclusively with equilibrium states, is not only inadequate for the description of complicated biological phenomena, it also fails to provide a sufficient description of dissipative physical and chemical phenomena. About 50 years ago, serious efforts were undertaken to extend the descriptive power of thermodynamics to systems that deviate from equilibrium by very small amounts—efforts primarily associated with the name of the Danish chemical engineer, Lars Onsager (1931).

Near-equilibrium, or irreversible thermodynamics as the subject came to be known, was predicated upon the assumption that one could treat state variables as scalar fields varying in space and time, in much the way temperature and pressure fields are portrayed in meteorology. To achieve this, one imagines the spatial domain of the system to be divided into a gridwork (not necessarily rectilinear) of sufficiently small cells. The cells are large enough to contain a macroscopic amount of material, but small enough so that any gradients of state variables can be reasonably approximated by a spatial succession of cells, each at thermodynamic equilibrium. Temporal variations within any segment must be slow. To satisfy these requirements, the spatial gradients must be gradual and the dissipation small, with the entire system deviating but little from thermodynamic equilibrium.

Gradients in state variables are observed to be accompanied by flow processes. For example, a spatial gradient in temperature is accompanied by a concomitant flux of heat energy (thermal diffusion); an analogous gradient in chemical potential gives rise to a diffusive material flux. The gradient need not always be in physical space—it may exist along some abstract dimension, such as the extent of reaction (i.e., how far a chemical reaction is from equilibrium). In this last case, if most of the chemical energy is possessed by the members of one side of a chemical reaction, those chemical species will react to drive the reaction towards equilibrium. In the general case, a gradient in a state variable is perceived as a "force" that induces a corresponding process ("flow") to occur. Near thermodynamic equilibrium, the fluxes are found to be empirically related to their forces in a linear fashion. Hence, if J_1 is an arbitrary flow and X_1 is its corresponding force, then near equilibrium

$$J_1 = L_{11}X_1 \tag{2.6}$$

where L_{11} is a "phenomenological constant." Examples of phenomenological "constants" are the Fourier coefficient of thermal conductivity or the Fickian diffusion constant. If, as in these examples, one already knows roughly what the driving force is, one may reexpress it in appropriate units so that the product of the thermodynamic force and its conjugate flux will have the dimensions of the rate of entropy production (energy/thermal degree/unit area or volume/time). In this manner, one may write

$$\sigma = J_1X_1 + J_2X_2 \tag{2.7}$$

for a system with two flows; or, in general,

$$\sigma = \sum_{i=1}^{n} J_i X_i \qquad (2.8)$$

for an n-process system, where σ is the rate of entropy production.

Natural processes rarely occur in isolation. For that matter, they rarely occur independently of each other. The interference of two processes is referred to as "coupling." Thus, when thermal conduction and mass diffusion occur in proximity, not only does each force induce its conjugate flow; each force is, in principle, capable of affecting nonconjugate flows as well. For example, subjecting a homogeneous fluid or solid mixture to an imposed thermal gradient not only results in a flow of heat through the medium, but it also will cause differential migration of one or more of the species in the mixture until the mass flux is balanced by an opposing gradient in the chemical potential(s) (Soret effect). Conversely, a diffusion of material gives rise to a thermal flux (Dufour effect). Perhaps more familiar examples of such coupling are the piezoelectric interaction of electrical and mechanical forces and flows and the thermocouple resulting from thermal and electrical interaction. To account for this coupling, the phenomenological relations must be modified. Near equilibrium, the linear expansion of flows in terms of the causal forces appears as

$$J_1 = L_{11}X_1 + L_{12}X_2$$
$$J_2 = L_{21}X_1 + L_{22}X_2 \qquad (2.9)$$

for a two-process system, or as

$$J_i = \sum_{j=1}^{n} L_{ij}X_j \qquad (2.10)$$

for a community of n processes.

Example 2.3

The Soret effect is fundamental to many mass separation techniques. For example, two bulbs are joined by an insulated tube of small diameter and initially filled with a mixture of two nearly ideal gases, for examples 90% hydrogen and 10% deuterium. (See Figure 2.3 and Example 18.5-2 in Bird et al., 1960.) One bulb is heated slightly to 375°K and the other is cooled to

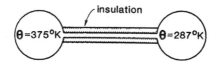

FIGURE 2.3. Schematic of an apparatus to separate gases using the Soret effect. Bulbs are maintained at different temperatures and the resulting heat flux is coupled to a differential mass flux.

287°K. The major heat flux down the tube will occur by thermal conduction. Fluxes of H_2 and D_2 will accompany the heat flux until balanced by a gradient in the concentrations of both gases. The coupling of the heat flux with both gases will be unequal, however, so that the more motile H_2 will preferentially migrate out of the hot bulb. When the system reaches steady-state, the hot bulb will contain about 89.6% H_2 and 10.4% D_2, whereas the cold bulb will contain relatively more (90.4%) H_2 and less (9.6%) D_2. The mixture in the hot bulb has been enriched in deuterium.

Lars Onsager (1931) showed that near equilibrium, the coupling coefficients were symmetrical with respect to forces and fluxes. For example, the coupling coefficient relating the thermal gradient to the mass flow in the Soret effect is equal in magnitude to that relating the gradient in chemical potential to the heat flux in the Dufour effect. In general, the result is that $L_{ij} = L_{ji}$ in (2.10).

In Onsager's statement on the symmetry of coupling, one finds intimations of the LeChâtelier-Braun principle, which says that any perturbation to a factor contributing to equilibrium induces a compensating change in an opposing factor. Thus, a disturbance in a temperature distribution not only causes a heat flow, but it also induces a mass flow. However, the symmetry of coupling insures that the induced gradient in chemical potential will serve to ameliorate the perturbing temperature gradient.

This relationship between the ideas of Onsager and LeChâtelier-Braun led Ilya Prigogine (1945) to restate the latter notion in more contemporary thermodynamic terms. Prigogine demonstrated that for an arbitrary ensemble of processes "sufficiently close to equilibrium states . . . the entropy production has its minimum value at the steady state compatible with the prescribed conditions." Substituting (2.9) into (2.7) and remembering that $L_{12} = L_{21}$ gives

$$\sigma = L_{11}X_1^2 + 2L_{12}X_1X_2 + L_{22}X_2^2 \tag{2.11}$$

as an expression for the rate of entropy production in a two-process system. In a general n-process community, the entropy production becomes

$$\sigma = \sum_{i=1}^{n} \sum_{j=1}^{n} L_{ij}X_iX_j. \tag{2.12}$$

In the neighborhood of thermodynamic equilibrium, the Prigogine theorem says that the steady-state configuration of forces will be such that (2.12) is the minimum possible under the constraints keeping the system from equilibrium. Because σ is a measure of the distance from thermodynamic equilibrium, the minimization of σ represents a return toward the unperturbed (equilibrium) state.

The Prigogine theorem has some severe restrictions, about which more will be said presently. It is very general, however, as it applies to an arbitrary suite of processes kept from equilibrium by very generalized constraints. To the biologist, however, what is most remarkable about the theorem is best exemplified by the phrasing many thermodynamicists give to the hypothesis when they are careless with their words; for example: "At steady-states near equilibrium the forces and fluxes *configure themselves* so as to minimize the rate of entropy production."

The Prigogine theorem is a variational principle. This means that there is an objective function, the value of which is maximized or minimized as the system approaches its final configuration. It is certainly not the first variational principle to be accepted into the physical sciences. Hamilton's principle in mechanics (Goldstein, 1950) is familiar to practically every student of the physical sciences. However, Prigogine deals with *whole* ensembles of processes. The various components of the whole system, the forces and fluxes, all coevolve so as to minimize the objective function, the entropy production.

Now the mathematical form used for expressing variational behavior is indistinguishable from that used to describe goal-seeking behavior. Given the taboo among many biologists against anything resembling teleology, it is not surprising that most are inclined to dismiss any proposition such as Prigogine's. But to use a given form of mathematics does not imply an opinion by the user regarding ultimate causes and ends. Mathematics is value-free. The phenomenological observation remains, however, that non-living, certainly non-thinking assemblages of rudimentary physical processes can interact in such a way as to optimize a common attribute; that is, they may behave like a *proto-community*. Thus, there appears to be no rational or empirical reason to prohibit the extension of such mathematical expression into the more complex biological world, teleology notwithstanding. From the pragmatic point of view, the variational calculus appears to be the most cogent way of quantitatively describing interacting processes. (There are other ways, as ecological modelers would be quick to point out, but more about that later.)

In spite of this exciting promise that Prigogine's theorem holds for whole system thinking, one should not lose sight of its limited applicability. For one thing, it is valid only very near to thermodynamic equilibrium, and it was discussed earlier how living systems are well removed from equilibrium. It has also been mentioned how skeptical one should be of any argument cast in terms of the entropy of a living system. However, Prigogine's arguments are all cast in terms of more readily measurable entropic flows, rather than the unquantifiable entropies, so the last criticism does not necessarily apply.

In the end, it is a pragmatic concern that poses the most serious obstacle to extending the hypothesis into biology. When considering physical processes, it seems a relatively easy (though certainly not trivial) task to identify the forces associated with observed flows. Temperature and pressure gradients and voltage differences or chemical affinities engender heat conduction, fluid flow, electrical current, and chemical reactions, respectively. But what are the forces behind the flows of biomass from grass to bison, hare to wolf, whale to remora, or lion to carrion? Of course, in specific instances, factors affecting these flows can be identified and sometimes quantified. But the identification of a generalized force in the thermodynamic sense has to date remained elusive (Ulanowicz, 1972). Efforts to define formal ecological forces in strict analogy with physical circumstances create more problems than they clarify (Smerage, 1976). If generalized ecological forces exist, they certainly remain obscure.

The ambiguity of forces in ecology and in much of the rest of biology is significant, because much of the effort spent trying to extend the Prigogine hypothesis beyond the near-equilibrium domain has proceeded on the assumption that generalized forces are palpable and can always be quantified (Glansdorff and Prigogine, 1971). If, indeed, the forces continue to elude the ecologist, what alternative is then left but somehow to circumvent the need for them? Matters can be placed in perspective by remembering that the purpose of thermodynamics is phenomenological description and not explanation, so that in this scheme, the forces (explanations) are of distinctly secondary importance. Other disciplines, notably the field of economics, have made reasonable progress in describing the behavior of systems using only flows without explicit mention of eliciting forces. It would appear a useful exercise, then, to review the study of fluxes and the networks they form in the search for suitable objects with which to describe ecosystems level phenomena.

2.5 Summary

Most of the confusion concerning the role of thermodynamics in science could be avoided by clarifying the phenomenological nature of the subject.

Thermodynamics is an effort to codify what is common to empirical observation. The power of thermodynamics lies in its generality, but this can be achieved only at the expense of deliberately ignoring lower level details. Reductionistic explanation, therefore, has no proper place in thermodynamics per se. Efforts, such as statistical mechanics, to interpret thermodynamic observations from microscopic mechanisms serve only to show that descriptions at the two levels are not inconsistent. Thermodynamics, however, is not contingent upon microscopic theories.

By and large humans directly observe the "microscopic" events in ecology; that is, the interactions between individual organisms. The lesson of statistical mechanics is that by aggregating events on the observable level, one might eventually infer some macroscopic (whole system) principle that becomes a significant addition to the body of thermodynamic laws. The mathematics of statistical mechanics do not appear to be sufficient for the purpose of making such inferences; a different calculus seems to be in order. It is important that any new discovery remains compatible with existing phenomenology. Therefore, these thermodynamic laws are scrutinized with an eye towards incorporating them into a later synthesis.

The first law of thermodynamics deals with both the conservation and the transformation of energy. Practically all existing descriptions of ecological energetics are cast in terms of conservation and could be encompassed by the old caloric theory. "Work," in the classical sense, does not figure prominently in ecological energetics. However, work may be viewed alternatively as an ordering process, and as such figures significantly in ecological processes. The second law can be interpreted as saying that it is impossible for any ecosystem component to convert all of its energetic and material inputs into ordered biomass (ecosystem work). Some fraction must always be dissipated.

The classical development of thermodynamics describes systems at thermodynamic equilibrium in terms of static or state variables. But living systems by definition do not exist at thermodynamic equilibrium. Therefore, the utility of state variables, such as entropy and free energy in ecosystems analysis, is open to question.

Efforts to extend thermodynamics beyond equilibrium treat processes as the result of putative "forces." When properly defined, the product of the forces and their conjugate flows has the dimensions of entropy production. A significant result is that any steady state system very near to equilibrium will always possess a configuration of forces and fluxes so that the rate of entropy production is a minimum. Although the mathematical form of this hypothesis resembles the treatment of goal-seeking behavior, the rudimentary nature of the systems it describes precludes any teleological interpretation. Therefore, the argument to prohibit the application of variational methods to more complex living systems on the grounds that such description is teleological cannot be valid. However, whatever variational statements that can now be made about living systems cannot explicitly involve known thermodynamic forces, because no one has yet identified a generalized ecological force. Thus, the most fruitful avenue appears to be to seek a variational statement expressed solely in terms of the network of flows.

3
The Object

"Παντα ρει"
(All is flux.)
Heraclitus ca. 500 B.C.

3.1 The Ubiquity of Flows

While most of science is concerned with the search for an explanation or cause for a given phenomenon, it is now clear that thermodynamics is a rare but powerful exception that is more concerned with the quantitative description of general phenomena. In this regard, then, irreversible thermodynamics since 1930 may be viewed as a departure from phenomenology, in that its formulators found it necessary to invoke forces (or explanations) to complete their descriptions of near-equilibrium events. But such forces are seen to be of marginal utility and generality. Forces at work in most real phenomena are obscure at best. It is difficult, for example, to describe the force that directs the mouse into the fox's jaws in any sense that has general application.

In contrast, the transformations and flows of matter and energy remain visible everywhere. Water in a teacup and matter in a galaxy are both observed to move at measurable rates. The rate of decay of neutrons into mesons and protons or that of Spartina leaves into dissolved organic material has been reasonably established. In fact, the early thermodynamicists were more preoccupied with process rates than with identifying forces. It was not until the latter part of the 19th Century that thermodynamic description began to digress from the quantitative narration of flow processes. One might even go so far as to argue that the generality of thermodynamic description is a consequence of the mind's ability to abstract the concept of flow from the infinite diversity and spectrum of real changes.

Although the observer may be ignorant of the details of what is happening in a complex system, some index of the rate of flow is usually extremely useful in assessing how well the system is functioning. The gross national product (a flow of currency) or the primary production of an estuary are direct examples of flow indices that characterize entire systems, whereas body temperature and heartbeat rate are indirect indications of biochemical fluxes and fluid advection within an organism.

Thus, if one describes an ecosystem in terms of material and energy flows, he makes the description in a language common to the narration of changes in the rest of nature. Then, should any pattern emerge from watching the development of ecological flows, it becomes a straightforward (although sometimes technically difficult) matter to check how ubiquitous that pattern is in the universe. If the flow description is sufficiently general, then one is potentially dealing with a new thermodynamic principle.

To be more specific, the pattern of development in ecosystems is said by some (Odum, 1971; Odum, 1977) to bear strong resemblance to patterns of development in economic, social, political, ontogenetic, cellular, and even meteorological systems (Boulding, 1978; Corning, 1983). As Stent suggested, maybe that pattern is easier to articulate in ecosystems than in the other disciplines, thereby affording ecologists the opportunity to advance the state of thermodynamic knowledge. Some quantitative scientists with a casual reading of ecology are wont to preach to the naturalist that ecology would progress significantly if only the ecologists would adopt more rigorous mathematical methods. While there is some truth in this criticism, the ecologist now has the opportunity to turn the tables, so to speak, by demonstrating that ecology may inspire significant advances in that centerpiece of physical science, thermodynamics. To achieve such progress, however, requires less of the transfer of mathematical techniques than it does the adoption of a sufficiently general perspective—one afforded by describing ecosystems in terms of material and energy flows.

Fortunately, mass and energy flows in ecosystems have been actively studied during the past 30 years. This is only natural, since ecology generally means the study of the relationships of organisms with one another and with their nonliving environment (Odum and Odum, 1959). Rather than fixing attention solely upon the organism or population itself, one should be equally concerned with what transpires between populations, and quantifying these transformations is most readily achieved by measuring the palpable fluxes.

Now, the thrust in thermodynamics towards generality runs counter to the direction of most biological description, which strives to differentiate life forms and processes to the greatest degree possible until each observation can be uniquely catalogued. The resulting picture is one of a living world so manifold, intricate, and complex as to seemingly defy any hope of uniform quantification. It is also, at times, a picture of extreme beauty; and anyone needlessly rejecting all detail does so at the peril of justifiably being labelled boorish and insensitive. Therefore, it is categorically stated here that this effort to portray ecosystem dynamics in terms of "brute" flows of material and energy (Engelberg and Boyarsky, 1979) is not an attempt to abnegate the value of descriptive biology or of matters numinous. Most will agree that an individual's creations in the arts or literature are valued more than his ability to consume food. Still, if the artist does not eat over a long enough period, he dies. Similarly, the members of an ecosystem may behave in bizarre and intricate ways in response to a myriad of nonmaterial cues in the environment. Most such behavior can be shown to affect the pattern of material and energetic exchanges in the ecosystem. Still, these extremely intriguing actions remain contingent upon the underlying networks of brute flows. Just as the phylogeneticist observes that those traits shared by the greatest number of species are the most primitive, so the phenomenological ecologist points out that those attributes shared by ecosystems with the rest of the universe are also the most elemental and, in at least one sense of the word, the most essential.

The flows make possible the higher level behaviors, which in turn help to order and coordinate the flows. So reflexive is this couple that the description of one member lies implicit in the description of the other element. It is in this reflexive sense that a key postulate in the development of the current thesis should be understood; to thermodynamically describe an ecosystem, it is sufficient to quantify the underlying networks of material and energy flows. A more general form of the postulate would read: *the networks of flows of energy and material provide a sufficient description of far from equilibrium systems.*

3.2 Describing Flow Networks

Thus far, the term "network" has appeared frequently without definition. Briefly, a *network* (or *graph*) is a collection of elements called *nodes*, pairs of which are joined to one another by a (usually larger) set of elements called *edges*. The nodes are ordered in some sequence, and an edge is identified by the names of the two nodes that it joins. An *arc* is an edge that possesses a sense indicated by the order in which the end nodes are specified; that is, its initial and terminal nodes. A *directed network* (or *digraph*) possesses only nodes and arcs. A *pathway* is a sequence of arcs, such that the terminal node of one arc is the initial node of the next arc in the sequence. In a *simple pathway*, no node is repeated twice. A *simple cycle* is a simple pathway, except that the terminal node of the last arc is identical to the initial node of the first arc. Networks with no simple cycles are called *acyclic*. If magnitudes are attached to each arc, the network is said to be *weighted*.

Example 3.1

The network of internal flows of energy (in kcal/m^2/y) occurring among the five components of the Cone Spring ecosystem (Tilly, 1968; Williams and Crouthamel, unpublished ms.) is depicted in Figure 3.1. It is a directed, weighted network. The sequence 1-2-3-4-5 traces a simple path-

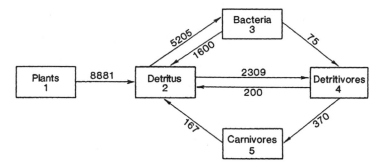

FIGURE 3.1. Schematic network of energy exchanges among the trophic compartments of the Cone Spring ecosystem (Tilly, 1968). Flows are measured in kcal $m^{-2}y^{-1}$.

way. Simple cycles are represented by the five sequences 2-3-4-5-2, 2-3-4-2, 2-4-5-2, 2-4-2, and 2-3-2.

Networks with all arcs originating and terminating at one of the nodes of the system are said to be *closed*. In *open* networks, at least some of the arcs originate or terminate outside the system (i.e., in the *universe* of thermodynamic parlance). Networks depicting living systems are always open, because living beings must exchange material and energy with their environment to survive.

Example 3.2

The completed Cone Spring network of energy transfers is depicted in Figure 3.2. Inputs to the system originate outside the network. The arrows pointing outwards (not terminating in a node) represent exports of energy in a form still useable to other systems; for example, energy incorporated into detrital material. At each node, the second law of thermodynamics requires that a certain amount of energy be dissipated; that is, lost to the system and unuseable by any other system at the same scale. These respirational flows are given the special ground symbol in Figure 3.2 and all succeeding diagrams.

In the life sciences, one is confronted with an entire hierarchy of networks. At any one level of observation, a node may be decomposed to yield an entire network at a lower level of the hierarchy. For example, the carnivores in Figure 3.2 may be resolved into species possibly having transfers among themselves. Each species in turn consists of individual organisms sometimes involved in a social structure. Each individual consists of organs or parts enmeshed in a coordinated exchange of energy and materials, etc. until one reaches the domain of molecular biology.

When one encounters examples of physical systems that grow or develop, such as a hurricane or a galaxy, there is an inclination to regard

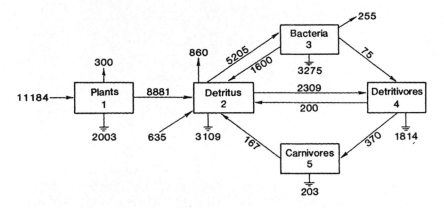

FIGURE 3.2. Schematic of the total suite of energy flows (kcal m^{-2}y^{-1}) occurring in the Cone Spring ecosystem. Arrows not originating from a box represent inputs from outside the system. Arrows not terminating in a compartment represent exports of useable energy out of the system. Ground symbols represent dissipations. Reprinted by permission of the publisher from Identifying the Structure of Cycling in Ecosystems, by R. E. Ulanowicz, Mathematical Biosciences vol. 65. Copyright 1983 by Elsevier Science Publishing Co., Inc.

these entities as continuua not readily amenable to description in terms of networks. While it is true that the continuum approach has heretofore proven very effective in fluid mechanics, this does not preclude one from viewing these phenomena as developing networks. For example, the spatial domain over which the organized flow occurs can be divided up into a gridwork of small spatial elements (an act that is now quite commonplace, given the popularity of numerical simulation techniques). Each spatial element becomes a node exchanging matter and energy with its immediate neighbors, thus giving rise to a flow network.

A nonbiological but living example of a flow network is the economy of a nation represented in terms of commodity flows among sectors. It was economic network analysis that gave rise to the quantitative techniques soon to be described.

Example 3.3

Economists usually categorize activity according to "sectors." The enumeration of sectors in a given economy can extend over all the corporate entities in a given nation, but most data are highly aggregated. In the most condensed representation, seven fundamental sectors are recognized: (1) agriculture, forestry, food, and tobacco; (2) energy; (3) mining, quarrying, and metals—excluding fossil fuels; (4) manufacture of equipment; (5) other manufacturing; (6) construction; and (7) trade, communications, and services. The commodities of foreign trade enter in the form of exogenous inputs and leave as useful exports. Consumer demand may be represented by the ground symbols used in ecosystems to signify respiration;

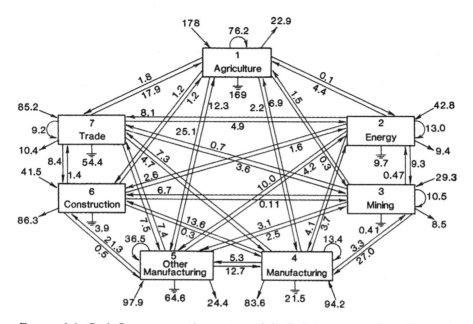

FIGURE 3.3. Cash flows among the sectors of the Polish economy (in millions of zlotys) during 1962. Exogenous inputs and outputs represent foreign trade. Ground symbols are used to represent consumer demand.

however, a strict analogy with dissipation is not implied by this notational convenience. Because the data are so condensed, the matrix of exchanges is highly connected.

Figure 3.3 portrays the currency exchanges that occurred in the Polish economy during the year 1962. The data were made available to the author by the Department of Regional Science at the University of Pennsylvania and were interpreted for the author by Janusz Szyrmer. All cash commodity flows are measured in millions of zlotys.

Finally, there is a host of other large-scale problems in organization that are readily described as networks—communication and transportation nets, information processing networks, and decision-making bodies. These structures all involve humans; and because they are perceived to share common evolutionary tendencies, the study of these various structures has recently begun to coalesce into the new and exciting discipline of "cognitive science." Teleology notwithstanding, there is reason to question why cognitive science should be limited to those systems involving subjects with consciousness. For just as engineering has been enriched by the study of functioning biological systems (bionics), might not social scientists likewise discern clues as to how social or political systems develop from studies conducted on nonthinking but developing entities such as ecosystems?

Returning to the field of ecology, one finds that the study of ecosystem flows has a recent albeit illustrious history. Eugene Odum (1953) discussed how ecological diversity might contribute to systems homeostasis by virtue of the associated multiplicity of flow pathways. A disturbance in any one pathway might be compensated by a change in some parallel chain of flow (cf., the LeChâtelier-Braun principle). MacArthur (1955) attempted to quantify the degree of parallel flow by borrowing an index from the then nascent field of information theory. Unfortunately, the focus soon shifted from *flow* diversity to *species* diversity, and the debate over the relationship of diversity to homeostasis floundered for the next 15 years. It has only been in the last few years that attention has begun to swing back to flow topology as a paramount descriptor of how ecosystems develop.

Besides its ostensible preoccupation with the brute aspects of ecosystem dynamics, two other objections are often raised against flow analysis. The first concerns the definition of the nodes of the network. Different individuals regarding the same ecosystem will inevitably choose different elements (nodes) with which to describe the transfers of material and energy. Some might insist on identifying a separate node for each species; others would aggregate many species into nodes representing trophic levels or large functional groups such as benthic detritivores. Most representations involve a compromise between fine resolution and a highly condensed system of a few elements. No single representation can be considered unique or absolute. How, then, does one draw robust conclusions from such subjective descriptions?

Indeed this "aggregation problem" (Halfon, 1979) does make life difficult when one's goal is the strict delineation of cause and effect, as it is with reductionistic modelling efforts. However, the phenomenological approach is not concerned with the detailed interactions. So long as the scale of description is large enough, a universal phenomenological law, such as one of the thermodynamic principles, will be independent of the level of the component nodes. The entropy of a galaxy will be observed to increase regardless of whether one identifies stars or moles of stellar gas as the nodes of interest. Similarly, one seeks here a principle of flow organization at the ecosystems level that will remain relatively insensitive to how the nodes of the network were chosen.

What turns out to be a more difficult criticism concerns whether enough different material fluxes have been measured to constitute a sufficient description of flows among the nodes. While there is only one network of energy exchanges for any set of ecosystem nodes at any given time, the number of fundamental material networks is limited only by the total number of chemical elements being transferred. Will all of the ecological phenomena be implicit in a network of carbon flows, or is it necessary also to measure the transfers of nitrogen, phosphorus, silicon, and other trace elements? Without delving into mechanisms, it is impossible to say a priori that any given flow description is complete. One can only state a systems level hypothesis and test whether the available flow data support the principle. If not, one should make an effort to gather data on other materials before rejecting the proposed law. Because working with several simultaneous networks causes mathematical complications, the assumption will temporarily be made that measurements of a single medium will be sufficient. Potential methods for treating multiple media will be discussed in Section 7.4.

Consideration of Example 3.2 reveals that four categories of flow appear in an ecological flow network: (1) flows among nodes within the system; (2) inputs of medium from outside the system; (3) exports of usable medium to other systems; and (4) exports of dissipated medium of no use to another system. For any n-component system, there are at most n^2 flows of the first type and n each of the remaining three kinds of flows. When the numbers of components and flows become large, pictorial representation of a network becomes cumbersome. Furthermore, analytical methods require a more abstract way of portraying the flow networks. For these reasons, it is useful to describe a network in terms of a single $n \times n$ square matrix and three n-element column vectors. (A brief review of matrix and vector manipulations is provided in Appendix A for those wishing to refresh their skill in this subject.)

Example 3.4

In the Cone Spring example (Figure 3.2), there are five nodes and eight internal transfers. Calling T_{ij} the transfer of energy from compartment i to compartment j, one sees that the T_{ij} constitute the elements of a 5×5 matrix

$$[T] \sim \begin{bmatrix} 0 & 8881 & 0 & 0 & 0 \\ 0 & 0 & 5205 & 2309 & 0 \\ 0 & 1600 & 0 & 75 & 0 \\ 0 & 200 & 0 & 0 & 370 \\ 0 & 167 & 0 & 0 & 0 \end{bmatrix}.$$

The external inputs to compartment i are denoted by D_i, the exports from compartment i by E_i, and the respirations by R_i. One is thus dealing with three n-compartment vectors,

$$(D) \sim \begin{pmatrix} 11,184 \\ 635 \\ 0 \\ 0 \\ 0 \end{pmatrix}, (E) \sim \begin{pmatrix} 300 \\ 860 \\ 255 \\ 0 \\ 0 \end{pmatrix}, (R) \sim \begin{pmatrix} 2003 \\ 3109 \\ 3275 \\ 1814 \\ 203 \end{pmatrix}.$$

The analysis of flows is facilitated when the system in question is at steady-state; that is, when the sum of all the inputs exactly balances the sum of all the outputs, or when

$$D_i + \sum_{j=1}^{n} T_{ji} = \sum_{k=1}^{n} T_{ik} + E_i + R_i \qquad (3.1)$$

for each node i. Systems not at steady-state may be treated by flow analysis (Hippe, 1983), but in this chapter, it will suffice to limit the discussion to balanced systems.

The description of flow topology should be in terms of quantities that are independent of the actual magnitudes of the flows. A convenient normalizing factor for each node is found by summing either the inputs or the outputs of the given node,

$$T_i' = D_i + \sum_{j=1}^{n} T_{ji} \qquad (3.2)$$

$$T_i = \sum_{k=1}^{n} T_{ik} + E_i + R_i \qquad (3.3)$$

where $T_i = T_i'$ at steady state. The T_i and T_i' are called the compartmental throughputs or throughflows and describe the level of flow activity through their respective compartments. Sometimes the throughputs bear little relation to the amount stored in a compartment, as best illustrated by the microflora in an ecosystem. The bacteria usually have a large throughput, although their stocks remain relatively small.

The size of the entire system (in terms of flow) is usually taken to be the sum of all the flows in the system

$$T = \sum_{j=1}^{n} \sum_{i=1}^{n} T_{ij} + \sum_{i=1}^{n} (E_i + R_i) + \sum_{j=1}^{n} D_j, \qquad (3.4)$$

where T is called the total system throughput. Another example of flows being used to measure the size of a system is the gross national product, or the sum of all the commodity inputs D_j to all sectors of the economy.

3.3 Analyzing Flow Networks*

Having established flow networks as the preferred focus of phenomenological attention and having sketched out the categories of flow to be measured, it is now possible to turn to the description of cycles and

---

* Sections 3.3 and 3.4 give helpful background to the synthesis to be presented in Chapter 6. It is not necessary, however, to be able to follow all the mathematical detail in these sections before one can understand the derivations occurring in that later chapter.

cybernetics in the next chapter. However, to proceed in such haste is certain to leave some readers feeling uneasy. Accustomed as everyone is to thinking in terms of strict cause and effect, it seems rather cavalier to dismiss the necessity for thermodynamic forces in such short order. What sense can be made out of a collection of flow values without any reference to what is keeping these flows in existence? Most will pose this question in microscopic fashion; for example, what conditions at nodes A and B cause the immediate flux from A to B to happen? However, it is the purpose of this book to convince the reader that questions about the long-term persistence of flows are better posed at the macroscopic level; that is, what contribution of a direct flow from A to B toward the configuration of the community of flows is propitious to the continued existence of the given flow?

However, even without such macroscopic perspective, it is still possible to make good sense out of a collection of directed flows. For example, knowing the topology of the flow network, and assuming the indistinguishability of quanta of medium, it becomes possible to estimate the fraction of the flow from A to B that had its origins in any of the several inputs to the system, or the fraction that is likely to leave the system at any given point. Although only direct flows are measured, indirect pathways from one node to another are implied by the network structure. These indirect flows can be assessed, and sometimes they are considerable (Patten, 1985). Furthermore, each flow may be perceived as a link in a trophic model. It is not surprising, therefore, that by studying the network structure, one may infer much about the trophic status of the communities. Some familiarity with the following methods of microscopic flow analysis should help assuage any lingering anxieties the reader may have concerning the assumption that flow networks are sufficient descriptors of ecosystem behavior.

Flow analysis is accomplished using the methods of linear algebra; that is, matrix and vector manipulations. The theory was first developed in the field of economics, which has traditionally been rich in data on commodity flows and depauperate in explanations of forces behind the flows. The so-called "input-output analysis" was first introduced into ecology by Hannon (1973).

The pivotal task in flow analysis is to relate the throughputs of each compartment to the exit flows from the system. Rearranging (3.3) to solve for the transfers outside the system gives

$$E_i + R_i = T_i - \sum_k T_{ik}. \tag{3.5}$$

This relationship may be recast in matrix-vector notation by defining the identity matrix $[I]$ with elements δ_{ik} ($\delta_{ik} = 0$ when $i \neq k$, and $\delta_{ik} = 1$ when $i = k$), and a matrix of partial "feeding" coefficients $[G]$ where the constituent element g_{ik} represents the fraction of the total input to k that comes directly from i (i.e., $g_{ik} = T_{ik}/T_k$). Then (3.5) becomes

$$E_i + R_i = \sum_k (\delta_{ik} - g_{ik})T_k, \qquad (3.6)$$

or

$$(E) + (R) = \{[I] - [G]\}(T), \qquad (3.7)$$

where (E), (R), and (T) are the column vectors with components E_i, R_i, and T_i, respectively. For the sake of brevity, the matrix inside the braces can be written simply as $[I - G]$, and it is referred to as the Leontief (1951) matrix. Solving (3.7) for the throughput in terms of the outputs gives

$$(T) = [I - G]^{-1}\{(E) + (R)\} \qquad (3.8)$$

where the superscript -1 indicates matrix inversion. The matrix $[I - G]^{-1}$ is called the input structure matrix (Hannon, 1973), or the Leontief inverse. It relates the activity of any compartment to the final exports and internal consumptions of the system.

Example 3.5

To calculate the throughput of any compartment in Cone Spring, one sums the corresponding column of the matrix $[T]$ in Example 3.4 and adds the appropriate component of the (D) vector in the same example. Alternatively, because the Cone Spring network is at steady-state, one could also sum the designated row in the $[T]$ matrix and add the corresponding components of (E) and (R).

$T_1' = 0 + 0 + 0 + 0 + 0 + 11,184 = T_1 = 0 + 8881 + 0 + 0 + 0$
$\quad + 300 + 2003 = 11,184$

$T_2' = 8881 + 0 + 1600 + 200 + 167 + 635 = T_2 = 0 + 0 + 5205$
$\quad + 2309 + 0 + 860 + 3109 = 11,483$

$T_3' = 0 + 5205 + 0 + 0 + 0 + 0 = T_3 = 0 + 1600 + 0 + 75 + 0$
$\quad + 225 + 3275 = 5205$

$T_4' = 0 + 2309 + 75 + 0 + 0 + 0 = T_4 = 0 + 200 + 0 + 0 + 370$
$\quad + 0 + 1814 = 2384$

$T_5' = 0 + 0 + 0 + 370 + 0 + 0 = T_5 = 0 + 167 + 0 + 0 + 0 + 0$
$\quad + 203 = 370$

The matrix $[G]$ is then calculated by dividing each component in any row of $[T]$ by its corresponding T_i.

$$[G] = \begin{bmatrix} 0 & .773 & 0 & 0 & 0 \\ 0 & 0 & 1 & .969 & 0 \\ 0 & .139 & 0 & .031 & 0 \\ 0 & .017 & 0 & 0 & 1 \\ 0 & .015 & 0 & 0 & 0 \end{bmatrix}.$$

One could have just as easily defined partial host coefficients as $f_{ij} = T_{ij}/T'_i$ and substituted into (3.2) to get

$$(T)' = [I - F^T]^{-1} (D) \tag{3.9}$$

where $[F]$ is a matrix with components f_{ij}, and the superscript after the $[F]$ matrix indicates matrix transposition. (Throughout this book, all vectors will appear as column vectors. Therefore, the well-known Augustinovics [1970] matrix $[I - F]$ appears in its transposed form, $[I - F^T]$.) Hence, the throughputs of each compartment also may be related to the external supplies. The conditions guaranteeing the existence of the Leontief (and Augustinovics) inverse are given by Hawkins and Simon (1949). The ways in which $[F]$ and $[G]$ have been normalized virtually insure that these conditions will be satisfied.

Knowing the output structure matrix $[I - F^T]^{-1}$, it now becomes a straightforward task to calculate the ultimate fate of any unit of input to the system. Consider for the moment that the appropriate unit vector replaces (D) in (3.9). (This analysis of flows per unit input is *intensive* in the sense described in Section 2.3.) The result of multiplying this unit vector on the left by the transpose of the Augustinovics inverse is the vector $(T)'$ of compartmental throughputs that would result from the single unit input. The accompanying internal exchanges T_{ij} are then calculated by denormalizing the $[F]$ matrix according to the $(T)'$ vector just calculated; that is, multiplying each f_{ij} by the respective T'_i as generated by the unit input. The sum of the exports and respirations may be determined by balance and then apportioned in the same ratio as they appear in the full network.

Example 3.6

To trace the fate of a single unit of detritus advected into Cone Spring, it is first necessary to calculate the matrix of partial host coefficients, $[F]$. This is done in similar fashion to the calculation of $[G]$ in the last example, only normalizing the *rows* of $[T]$ by their respective T_i's (Patten et al., 1976).

$$[F] = \begin{bmatrix} 0 & .794 & 0 & 0 & 0 \\ 0 & 0 & .453 & .201 & 0 \\ 0 & .307 & 0 & .014 & 0 \\ 0 & .084 & 0 & 0 & .155 \\ 0 & .451 & 0 & 0 & 0 \end{bmatrix}.$$

Using $[F]$ to calculate the transpose of the Augustinovics inverse,

$$[I - F^T]^{-1} = \begin{bmatrix} 1.000 & 0 & 0 & 0 & 0 \\ .958 & 1.210 & .374 & .186 & .545 \\ .434 & .547 & 1.170 & .084 & .247 \\ .199 & .251 & .092 & 1.040 & .113 \\ .031 & .039 & .014 & .161 & 1.020 \end{bmatrix}.$$

Multiplying this output structure matrix by the second unit vector (i.e., a column vector with one in the second position and zero elsewhere) yields the throughputs generated by the unit input. These throughputs are readily seen to be equal to the second column vector of the output structure matrix,

$$(T)_2' = \begin{pmatrix} 0 \\ 1.210 \\ .547 \\ .251 \\ .039 \end{pmatrix}.$$

The exchange matrix generated by this unit input subsequently is found by multiplying each row of the host coefficient matrix $[F]$ by its corresponding throughput in $(T)_2'$.

$$[T] = \begin{bmatrix} 0 & 0 & 0 & 0 & 0 \\ 0 & 0 & .547 & .243 & 0 \\ 0 & .168 & 0 & .008 & 0 \\ 0 & .021 & 0 & 0 & .039 \\ 0 & .018 & 0 & 0 & 0 \end{bmatrix}.$$

The sum of export and respiration from each component is determined by difference. For example, there is only one internal input to compartment 3 (bacteria) coming from compartment 2 (detritus) in the amount of .547 units. Two outputs issue from the bacteria to other compartments: .168

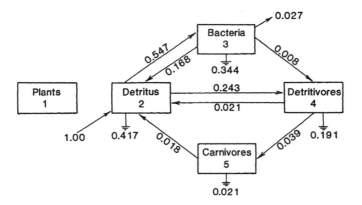

FIGURE 3.4. The fate of a single arbitrary unit of energy accompanying the detritus imported into the Cone Spring ecosystem.

units to the detritus and .008 units to the detritus feeders. By difference, .371 units are either respired or exported. Because about 7% of total exports in the original network were useable and the remaining 93% passed from the system as respiration, one estimates that .344 units are respired and .027 units are advected out of the system. The fate of the unit of imported detritus is schematized in Figure 3.4.

Once the fate of each of the separate inputs is determined, it becomes an easy matter to assess the proportions of any internal exchanges that are attributable to the separate inputs under the assumption of uniform mixing. For instance, in Example 3.6, it was determined that the import of energy associated with the detritus at the rate of 1 kcal m^{-2} y^{-1} gave rise to carnivory on the detritus feeders of .039 kcal m^{-2} y^{-1}. If one repeats the analysis in Example 3.6 for a unit of primary productivity, one finds that .0309 kcal m^{-2} y^{-1} of the detritus feeders will be consumed by the carnivores. Since 635 kcal m^{-2} y^{-1} of detritus actually enter the system and 11,184 kcal m^{-2} y^{-1} of primary production were measured, one estimates that of the 370 kcal m^{-2} y^{-1} flowing from detritivores to carnivores, approximately 25 units trace their origin to imported detritus and 345 to the plant production.

In short, one may use the structure matrices to analyze any direct flow within the system. One must keep in mind, however, that there are linear assumptions implicit in input-output analysis. Because it is possible to estimate the origins and fates of any given flow in a static network, it does not follow that a change in that flow (or its sources and sinks) will be distributed according to the principle of linear superposition.

Thus far, attention has been focused on the origins and fates of direct flows. However, the same structure matrices also contain information on the magnitudes of all indirect flows occurring between any two components of the system.

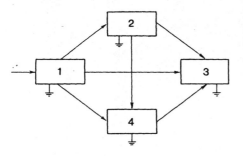

FIGURE 3.5. The flow connections among the four compartments of a hypothetical network.

To see this, it is useful to consider the significance of the powers of the $[F]$ and $[G]$ matrices. The components of the matrix $[F]$ represent the fraction of flow through compartment i (the row index), which proceeds directly to compartment j (the column index). If one multiplies $[F]$ by itself, the result is denoted by $[F]^2$. The i-j^{th} component of $[F]^2$ represents the fraction of the total flow through i, which flows into j along all pathways of exactly two transfers. Similarly, $[F]^m$ will have an i-j component that represents the fraction of T_i, which flows to j along all pathways of exactly m links.

Example 3.7

The $[F]$ matrix for the simple network in Figure 3.5 looks like

$$[F] = \begin{bmatrix} 0 & f_{12} & f_{13} & f_{14} \\ 0 & 0 & f_{23} & f_{24} \\ 0 & 0 & 0 & 0 \\ 0 & 0 & f_{43} & 0 \end{bmatrix},$$

which when multiplied by itself yields

$$[F]^2 = \begin{bmatrix} 0 & 0 & (f_{12}f_{23} + f_{14}f_{43}) & f_{12}f_{24} \\ 0 & 0 & f_{24}f_{43} & 0 \\ 0 & 0 & 0 & 0 \\ 0 & 0 & 0 & 0 \end{bmatrix}.$$

Inspection of Figure 3.5 confirms that there are exactly four pathways of length 2 in the network. Two of these begin at 1 and end at 3 (accounting for the two terms in the 1–3 component of $[F]^2$). When $[F]^2$ is multiplied (on the left) by $[F]$ one more time, the result is

$$[F]^3 = \begin{bmatrix} 0 & 0 & f_{12}f_{24}f_{43} & 0 \\ 0 & 0 & 0 & 0 \\ 0 & 0 & 0 & 0 \\ 0 & 0 & 0 & 0 \end{bmatrix}.$$

Again, inspection confirms that there is only one pathway of length 3 in the network, as indicated by the only nonzero component, 1–3. That pathway starts at 1 and proceeds via 2 and 4 before terminating at 3. Higher powers of $[F]$ are identically zero, implying that no pathways with a length greater than 3 are present in the network. If there are no cycles in the network, the powers of $[F]$ will always truncate prior to reaching $[F]^n$, where n is the number of nodes in the network. If cycles are present in the network, the powers of $[F]$ form an infinite sequence.

If one now asks the question, "What is the total fraction of T_i that flows to compartment j along all pathways of all lengths?," the answer is obtained by summing all the powers of $[F]$, that is,

$$\sum_{m=1}^{\infty} [F]^m = [F] + [F]^2 + [F]^3 + \ldots \qquad (3.10)$$

Now the components of $[F]$ are defined so that each $f_{ij} \le 1$. The infinite series (3.10), therefore, may possibly converge to a finite limit. In fact, if one adds the identity matrix $[I]$ (or, equivalently, $[F]^0$) to series (3.10), the limit may be expressed as $[I - F]^{-1}$, that is,

$$[I - F]^{-1} = \sum_{m=0}^{\infty} [F]^m. \qquad (3.11)$$

But this limit is nothing more than the Augustinovics inverse.

Thus, if one subtracts unity from each diagonal element of the output structure matrix, the i-j component of the resulting matrix represents exactly the fraction of T_j flowing to i over *all* possible pathways.

A parallel argument leads to the conclusion that

$$[I - G]^{-1} = \sum_{m=0}^{\infty} [G]^m. \qquad (3.12)$$

Thus, after 1 is subtracted from the diagonals, the i-jth component of the input structure matrix represents the fraction of T_j, which is dependent upon T_i via all pathways of all lengths.

Example 3.8

Describe the indirect effects inherent in the hypothetical network of Figure 3.6. The $[G]$ matrix for Figure 3.6 is

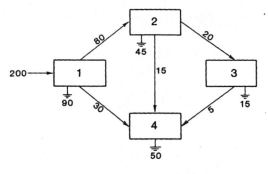

FIGURE 3.6. A hypothetical network of flows among a four-compartment system. Units are arbitrary.

$$[G] = \begin{bmatrix} 0 & 1 & 0 & .6 \\ 0 & 0 & 1 & .3 \\ 0 & 0 & 0 & .1 \\ 0 & 0 & 0 & 0 \end{bmatrix}.$$

So that $[I - G]^{-1}$ becomes

$$\begin{bmatrix} 1 & 1 & 1 & 1 \\ 0 & 1 & 1 & .4 \\ 0 & 0 & 1 & .1 \\ 0 & 0 & 0 & 1 \end{bmatrix}.$$

Subtracting 1 from the diagonals gives

$$\begin{bmatrix} 0 & 1 & 1 & 1 \\ 0 & 0 & 1 & .4 \\ 0 & 0 & 0 & .1 \\ 0 & 0 & 0 & 0 \end{bmatrix}.$$

Now the 1–2 component is unity, indicating that 2 is completely dependent upon 1 for its input. Similarly, 3 is completely dependent upon both 1 and 2 for its inputs. Although 4 receives only 60% of its input directly from 1, it is obvious that 4 is completely dependent upon 1 for its input, since the other two inputs to 4 are directly or indirectly dependent on 1. This results in a value of 1 in the 1–4 component of the $[I - G]^{-1}$ matrix. Both the 2–4 and 3–4 flows into 4 are dependent upon the existence of 2, so that 40% of the input to 4 is dependent upon 2, as shown in the 2–4 component of the structure matrix. Finally, 10% of the throughput of 4 is contingent upon 3.

When cycles are present in the network, it is possible for the diagonal components of the Leontief inverse to exceed unity, and this numerical feature will be used to good advantage in Section 4.3.

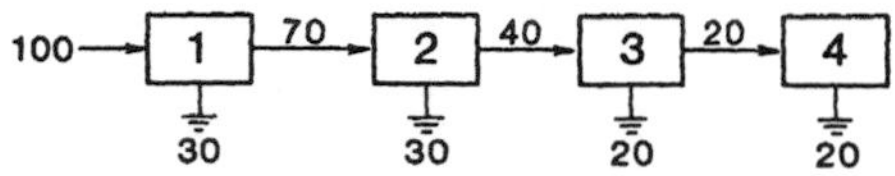

FIGURE 3.7. Flows along an imaginary trophic chain.

A curious thing happens when one looks at the input structure matrix of a straight chain of transfers, such as the one in Figure 3.7. It is obvious that the last member of this chain is dependent upon all the previous members. It is not surprising, therefore, that when one calculates the Leontief inverse for the straight chain, the last column has 1's in all its entries. The sum of the entries in the last column equals the trophic position of the last member of the chain. In fact, the sum of each column in the matrix is equal to the trophic position of the corresponding compartment.

$$[I - G]^{-1} = \begin{bmatrix} 1 & 1 & 1 & 1 \\ 0 & 1 & 1 & 1 \\ 0 & 0 & 1 & 1 \\ 0 & 0 & 0 & 1 \end{bmatrix}$$

(3.13)

column sums 1 2 3 4

Comparing the $[I - G]^{-1}$ matrix from this straight chain with its counterpart in Example 3.8, one finds that the first three columns of both structure matrices are identical. The column sum of the fourth compartment in Example 3.8 is lower (2.5) than its counterpart in the straight chain, reflecting the subsidies that this compartment receives from lower trophic level components. This smaller value is nonetheless quite reasonable. Compartment 4 of Example 3.8 receives 60% of its flow from the first trophic level (i.e. it acts 60% like a herbivore at level 2), 30% from the second, and 10% from the third. Its equivalent trophic position is thus $.6 \times 2 + .3 \times 3 + .1 \times 4 = 2.5$.

It is no accident that the column sums of the input structure matrix reflect the trophic positions of their corresponding network nodes. Levine (1980) has shown rigorously that a column sum of the Leontief inverse is precisely the equivalent trophic position of the corresponding component. (Here the word "trophic" is used broadly to include processes such as detritivory, saprophagy, etc.) This allows one to assign equivalent trophic positions to species embedded in arbitrarily complex networks. Significant changes in the values of the equivalent trophic positions, or in their relative rankings, often are indicators of ecosystem response to stress.

Example 3.9

To determine the equivalent trophic level for each compartment in Cone Spring, one begins with the [G] matrix as calculated in Example 3.5.

$$[G] = \begin{bmatrix} 0 & .773 & 0 & 0 & 0 \\ 0 & 0 & 1 & .969 & 0 \\ 0 & .139 & 0 & .031 & 0 \\ 0 & .017 & 0 & 0 & 1 \\ 0 & .015 & 0 & 0 & 0 \end{bmatrix}$$

Matrix manipulation gives the Leontief inverse as

$$[I - G]^{-1} = \begin{bmatrix} 1 & .933 & .933 & .933 & .933 \\ 0 & 1.210 & 1.210 & 1.210 & 1.210 \\ 0 & .169 & 1.170 & .201 & .201 \\ 0 & .039 & .039 & 1.040 & 1.040 \\ 0 & .018 & .018 & .018 & 1.020 \end{bmatrix}$$

and summing the columns gives the effective trophic positions—1 for the plants (as expected), 2.37 for the detritus, 3.37 for the bacteria, 3.4 for the detritus feeders, and 4.4 for the carnivores. (These trophic values are incremented beyond their expected integer values because of the presence of cycled flow in the system.)

Levine used the $[G]$ matrix to distribute the trophic levels among the species. However, it is also possible to invert this process and distribute the species among distinct trophic levels (Ulanowicz and Kemp, 1979). This allows one to think of the network as being mapped into a concatenated series of trophic levels in the sense of Lindeman (1942).

To see how aggregate trophic levels are defined, one begins by dividing the magnitude of each external input by the throughput of the receiving compartment to give the fraction of the particular species that acts as a "primary producer," $\delta_i = D_i/T_i$. One assigns this fraction to the first trophic level. Call (d_1) a vector with elements δ_i. Now, multiplying (d_1) on the left by $[G^T]$ yields the fraction of this newly introduced medium, which each species receives along all single-length pathways; that is, that fraction of T_i being supplied at the second trophic level. Call the vector of these fractions (d_2) so that

$$(d_2) = [G^T] (d_1). \tag{3.14}$$

Similarly, the fraction of each species that feeds at the third trophic level is

$$(d_3) = [G^T]^2 (d_1), \tag{3.15}$$

while the fraction that feeds at the mth trophic level is

$$(d_m) = [G^T]^{m-1} (d_1). \tag{3.16}$$

If no cycles exist in the network, the powers of $[G]$ truncate after $n - 1$ or fewer steps. Even if cycles do exist, the (d_m) become vanishingly small as m increases, so that one may summarily truncate the series after a conveniently large m. Fortunately, the average number of direct trophic transfers in ecosystems is usually quite small (Pimm and Lawton, 1977).

It is now possible to define an $m \times n$ dimensional trophic transformation matrix $[M]$, in which the ith *row* is taken to be $(d_i)^T$. Equation (3.1) may be written in matrix-vector form as

$$(D) + [T]^T(1) = [T](1) + (E) + (R), \tag{3.17}$$

where (1) is an n-dimensional column vector of ones, and $[T]$ is the matrix of intercompartmental transfers defined earlier. Thus, multiplying (3.17) on the left by $[M]$ yields,

$$[M](D) + [M][T]^T(1) = [M][T](1) + [M](E) + [M](R). \tag{3.18}$$

But it may be shown from preceding equations that for any acyclical network

$$[M]^T(1)^* = (1), \tag{3.19}$$

where $(1)^*$, is an m-dimensional column vector of 1's. Equation (3.19) is also approximately true for cyclical networks when m has been chosen conveniently large. Equation (3.18) may therefore be rewritten as the *m-dimensional* system

$$(\mathcal{I}) + [\mathcal{T}]^T(1)^* = [\mathcal{T}](1)^* + (\mathcal{E}) + (\mathcal{R}), \tag{3.20}$$

where
$$[\mathcal{T}] = [M][T][M]^T \tag{3.21}$$

$$(\mathcal{I}) = [M](D) \tag{3.22}$$

$$(\mathcal{E}) = [M](E) \tag{3.23}$$

and
$$(\mathcal{R}) = [M](R). \tag{3.24}$$

Because matrix $[\mathcal{T}]$ is not unity on the diagonal above the primary diagonal and zero elsewhere, (3.20) does not represent a strict straight chain system. However, because of a peculiar symmetry in $[\mathcal{T}]$, an equivalent concatenated chain can be constructed using (3.20). The input to the first trophic level is assumed to be the sum of all the components of $(\mathcal{I})$. The output from trophic level i to the next trophic level is taken to be the sum of the ith row of $[\mathcal{T}]$. The ith components of $(\mathcal{E})$ and $(\mathcal{R})$ will finish the balance of the ith compartment.

Example 3.10

It is desired to compute the trophic transformation matrix $[M]$ for the network in Figure 3.6. One begins with the transpose of the $[G]$ matrix as shown in Example 3.8:

$$[G^T] = \begin{bmatrix} 0 & 0 & 0 & 0 \\ 1 & 0 & 0 & 0 \\ 0 & 1 & 0 & 0 \\ .6 & .3 & .1 & 0 \end{bmatrix},$$

and with the normalized input vector

$$(d_1) = \begin{pmatrix} 1 \\ 0 \\ 0 \\ 0 \end{pmatrix}.$$

The transpose of (d_1) becomes the first row of $[M]$. To calculate the second row of $[M]$, one premultiplies (d_1) by $[G^T]$ to get

$$(d_2) = [G^T]\,(d_1) = \begin{pmatrix} 0 \\ 1 \\ 0 \\ .6 \end{pmatrix},$$

and further

$$(d_3) = [G^T]\,(d_2) = \begin{pmatrix} 0 \\ 0 \\ 1 \\ .3 \end{pmatrix},$$

and finally

$$(d_4) = [G^T]\,(d_3) = \begin{pmatrix} 0 \\ 0 \\ 0 \\ .1 \end{pmatrix}.$$

Whence, the $[M]$ matrix becomes

$$[M] = \begin{bmatrix} 1 & 0 & 0 & 0 \\ 0 & 1 & 0 & .6 \\ 0 & 0 & 1 & .3 \\ 0 & 0 & 0 & .1 \end{bmatrix}.$$

To describe the composition of the trophic levels, one reads across the rows. Thus, the third trophic level consists of all of compartment 3 and 30% of compartment 4. To see how the compartments are assigned to the trophic levels, one reads down the columns. Hence, 60% of compartment 4 is assigned to trophic level 2, 30% to level 3, and 10% to level 4.

An equivalent trophic straight chain is assembled using $[M]$. The inputs to each member in the chain come from premultiplying the vector of throughputs (T) by $[M]$

$$\begin{bmatrix} 1 & 0 & 0 & 0 \\ 0 & 1 & 0 & .6 \\ 0 & 0 & 1 & .3 \\ 0 & 0 & 0 & .1 \end{bmatrix} \begin{pmatrix} 200 \\ 80 \\ 20 \\ 50 \end{pmatrix} = \begin{pmatrix} 200 \\ 110 \\ 35 \\ 5 \end{pmatrix}.$$

The respirations are the result of premultiplication of (R) by $[M]$.

$$(\mathfrak{R}) = [M](R) = \begin{pmatrix} 90 \\ 75 \\ 30 \\ 5 \end{pmatrix}.$$

There are no exports in this case, but their canonical forms would otherwise be calculated exactly like the respirations. The ensuing straight chain is shown in Figure 3.8.

Example 3.11

Steele (1974) estimated the energy flow among 10 compartments of the North Sea ecosystem. The trophic transformation matrix corresponding to Figure 3.9 is calculated as in Example 3.10,

$$[M] = \begin{bmatrix} 1 & 0 & 0 & 0 & 0 & 0 & 0 & 0 & 0 & 0 \\ 0 & 1 & 0 & 0 & 0 & 0 & 0 & 0 & 0 & 0 \\ 0 & 0 & 1 & .885 & 1 & 0 & 0 & 0 & 0 & 0 \\ 0 & 0 & 0 & .115 & 0 & .099 & .833 & 1 & 0 & 0 \\ 0 & 0 & 0 & 0 & 0 & .707 & .167 & 0 & .833 & .099 \\ 0 & 0 & 0 & 0 & 0 & .185 & 0 & 0 & .167 & .707 \\ 0 & 0 & 0 & 0 & 0 & .009 & 0 & 0 & 0 & .185 \\ 0 & 0 & 0 & 0 & 0 & 0 & 0 & 0 & 0 & .009 \\ 0 & 0 & 0 & 0 & 0 & 0 & 0 & 0 & 0 & 0 \\ 0 & 0 & 0 & 0 & 0 & 0 & 0 & 0 & 0 & 0 \end{bmatrix}$$

and results in the equivalent straight chain of Figure 3.10.

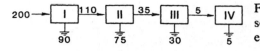

FIGURE 3.8. A straight chain representation of the trophic flows inherent in the network of Figure 3.6.

Finally, it is possible to perform a sensitivity analysis on the structure matrices to estimate how an infinitesimal change in any one compartment or flow would affect the remainder of the network. Through such analysis, it is possible to identify potentially controlling links in a network. For details of the sensitivity methods, the reader is referred to Bosserman (1981).

3.4 Standing Stocks and Fluxes

The many benefits provided by input-output analysis might still fail to soothe all of the reader's anxieties about the postulate of flow sufficiency. Certainly, there is no lack of critics of economic input-output analysis (Georgescu-Roegen, 1971). The chief concern seems not to be the neglect of forces, but rather the conviction that certain economic events occur

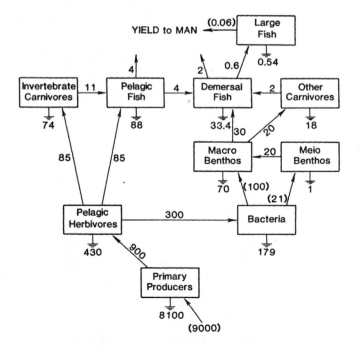

FIGURE 3.9. Estimated flows (kcal m^{-2}y^{-1}) through the food web of the North Sea (after Steele, 1974). Reprinted by permission of Harvard University Press.

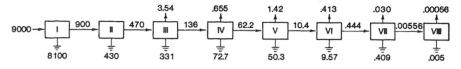

FIGURE 3.10. A straight chain of trophic transfers inherent in the North Sea food web (Figure 3.9). All flows are in kcal $m^{-2}y^{-1}$.

primarily because there is a collection of capital in certain sectors, and the flow analysis presented here does not consider such standing stocks.

It is possible to address these concerns in either of two ways. The first, patterned after leads made by Patten (1982), effectively blurs the distinction between stocks and flows. The compartments at time t are considered as distinct from those at some later time, $t + \theta$. Over the finite interval θ, a flux per unit time, T_{ij}, results in the transfer of θT_{ij} units of medium from i to j. A certain amount (S_i) of medium in i at time t remains in i at $t + \theta$. This standing stock over the interval θ is added to the diagonal element θT_{ii} in the matrix of total amounts transferred among compartments $[T]_\theta$. By normalizing either the rows or columns of the transfer matrix, one obtains matrices $[F]_\theta$ and $[G]_\theta$, respectively. It then becomes possible to perform a full input-output analysis for any interval θ. If the system is at steady-state, the matrices $[F]_\theta$ and $[G]_\theta$ will approach $[F]$ and $[G]$ as time goes to infinity. Effectively, a standing stock has been treated as a transfer-in-place.

Example 3.12

The elementary two component network in Figure 3.11 is at steady-state. The observed standing stocks are 500 units in component 1 and 200 in 2. The matrix of internal flows and $[G]$ look like:

$$[T] = \begin{bmatrix} 0 & 50 \\ 10 & 0 \end{bmatrix}, \qquad [G] = \begin{bmatrix} 0 & 1 \\ .091 & 0 \end{bmatrix}.$$

If the flows are accumulated for 3 units of time, it is evident that during this interval, 150 units will flow directly from 1 to 2 and 30 units from 2 to 1. If one approximates the flow dynamics as linear, then about 258 of the original 500 units of 1 will remain in 1 after the interval and 94 of the 200

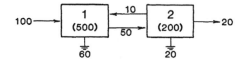

FIGURE 3.11. A rudimentary network of mass flows between two imaginary system elements. Flows have the units of mass/time, while the standing stocks of mass are indicated within the boxes.

units in 2 will stay in place. The total transfer matrix and the corresponding $[G]_\theta$ for the interval $\theta = 3$ then become:

$$[T]_{\theta=3} = \begin{bmatrix} 258 & 150 \\ 30 & 94 \end{bmatrix}, \qquad [G]_{\theta=3} = \begin{bmatrix} .439 & .615 \\ .051 & .385 \end{bmatrix}.$$

It is left as an exercise for the reader to show that as $\theta \to 0$, $[G]_\theta \to [I]$, and as $\theta \to \infty$, $[G]_\theta \to [G]$.

The second, and perhaps preferred way of accounting for the influence of standing stocks is to describe their consequences as they appear in the time history of flows. This treatment of dynamic networks is deferred until the last chapter. Suffice it to say here that a standing stock has the effect of altering the phase relationships between the inputs and outputs of a compartment. Therefore, the effects of standing stocks will be implicit in the description of the phase relationships among the various flows in the network.

3.5 Summary

Processes or flows are more fundamental and perceptible than the more recent concept of thermodynamic forces. They were also more important in the early development of thermodynamics than is apparent from reading most textbooks. Flows are a phenomenon common to practically every scientific discipline, so that most objects of scientific study can be described in terms of flux networks. A universal principle of growth and development thus might take the form of describing the evolution of flow networks.

Flows and transfers are implicit in the very definition of ecology. Because much effort has been expended recently in articulating flow processes in ecosystems, it is conceivable that ecology will provide the lead in formulating a universal theory of development in natural networks.

Flows are necessary for the maintenance of all life. In the strict sense, they are not sufficient to describe all aspects of growth and development in living systems. However, feedback between the higher aspects of life forms (e.g., ethology) and the more elemental transfers of material and energy is usually quite strong, so that the more complicated phenomena are likely to be implicit in the description of flow networks. The working hypothesis is therefore made that networks of energy and material flows provide a sufficient phenomenological description of growth and development in ecological and other far-from-equilibrium systems.

Flows occurring in ecosystems may be portrayed as directed, weighted graphs composed of elemental simple pathways and simple cycles. Four categories of flows may be identified in ecosystems: (1) inputs from outside the system; (2) transfers among components within the system; (3)

exports of useable medium from the system; and (4) dissipation of medium into a nonuseful form. The same categorization pertains to flows at all levels of the biological hierarchy, from organelles to the biosphere. Furthermore, a host of nonbiological systems are amenable to description in the same terms—economic entities, fluid dynamic and meteorological complexes, communication networks (both human and nonhuman), transportation grids, information processing networks, and distributed decision-making bodies such as governments.

The flows of a network may be written as the components of matrices and vectors with dimensions equal to the number of nodes in the network. Thus, one may use linear algebra to investigate many of the microscopic properties of steady-state networks: (1) the relationship between any two flows in the system may be quantified using the established methods of economic input-output analysis; (2) the effective trophic level of each species is readily quantified; (3) the species may each be mapped into the members of a discrete, concatenated trophic chain; and (4) linear sensitivity analysis may be applied to locate both the most vulnerable and the most controlling links in the network.

Although some criticize flow analysis for neglecting standing stocks, the effects of these pools of medium are seen to be implicit in suitable flow descriptions of either steady-state balance or dynamic change.

4
An Agent

> ". . . the cause of the event is neither in the one nor in the
> other but in the union of the two.
>   Or in other words, the conception of a cause is inapplicable
> to the phenomena we are examining."
>
> Leo Tolstoy
> *War and Peace*, Epilogue II

4.1 Cycles and Autonomous Behavior

There was almost no mention in the last chapter about cycles as they
occur in living networks. This is not to deem them inconsequential; on the
contrary, they deserve special consideration as agents capable of strongly
influencing the overall network structure. However, to appreciate this
potential, it becomes necessary to regard a flow cycle as a structure with
an existence that is to some degree independent of its constituents. To
show that such autonomy is a strong probability becomes a rather delicate
task involving reflexive reasoning that bears resemblence to circular
logic. Of course, if nature abhors a vacuum, then it might as readily be
said that logic abhors a cycle. Most arguments are structured to proceed
over an unambiguous chain of reasoning from antecedent to consequence,
from cause to effect, from force to ensuing transition. Circular reasoning
is generally to be avoided. An effect as its own cause is unsettling to
almost everyone. Hence, anyone addressing circular causality in nature is
almost certain to be suspect, if not for his motives, then at least for
weaknesses in his reasoning.

How, then, does one go about describing a circular configuration of
events? If one attempts a description of the whole configuration, it becomes all too easy to commit errors. It is far easier to avoid causal cycles
and to focus instead on a simpler subsystem. For example, Figure 4.1 is a
schematic of a perfect causal cycle—B is a direct consequence of A, C of
B, D of C and finally A is a consequence of D (Hutchinson, 1948). Entity A
becomes wholly a consequence of itself at an earlier time—a difficult
situation, to be sure. It is much safer to regard the component pathway of
Figure 4.2. Now B becomes the primal cause, D the ultimate effect, and
everything is straightforward.

Alternatively, one might analyze a particular element of the cycle, C for
example, in terms of its component pieces. Or possibly a single bilateral
relation could be described in detail; for example, the transition from B to
C. This is what is most often done in ecological modelling, where the

FIGURE 4.1. An imaginary closed causal cycle.

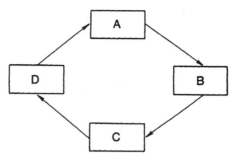

arrow from B to C might represent, for example, a flow of nitrogen. If one can quantitatively describe the flux from B to C in terms of some proximate causes, then one maintains a hope of describing the larger system. For example, the feeding behavior of C on B combined with a measure of the availability of B might give rise to a Lotka-Volterra or Michaelis-Menten functional description. The usual strategy consists of exhaustively describing all bilateral interactions, combining them according to some balance scheme (usually the conservation of energy or mass), simulating these combined interactions, and hoping the simulation describes how the aggregated system behaves. If the description is poor, the tendency is usually to seek a more refined (i.e., more reductionistic) model.

The peril in these strictly reductionistic descriptions is that one may completely lose sight of any autonomous attributes the cycle (as a unit) might possess. Autonomous behavior cannot be traced to any external antecedents. Formally, one may regard the causal loop of Figure 4.1 as wholly autonomous. There are no exogenous inputs (causes or forces) to the system. Break the loop in any way (as in Figure 4.2) and one observes a decidedly nonautonomous subsystem.

"But autonomy is an illusion," the reader might object, "Microscopically perturb any one of the constituent parts and the circulation of causality is sure to be affected." This is true, at least in the very short term. However, on a longer time scale, the converse of this objection could be equally true—the very existence of the cycle will affect the composition of the constituent parts. Suppose, for example, the cycle in Figure 4.1 exhibits what is known as positive feedback; that is, an increase in the flow from A to B will engender an increase in the flow from B to C, which in turn increases C to D, etc. (e.g., "hypercycles" of Eigen, 1971). A chance perturbation in any constituent element that serves to increase a constituent flow will be self-reinforcing. Any chance happening that decreases a constituent flow will receive no reinforcement at all. This rat-

FIGURE 4.2. A linear causal chain.

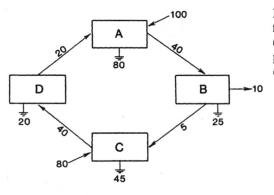

FIGURE 4.3. A hypothetical, but feasible, cycle of material flows (in arbitrary units)·with accompanying gains and losses at each node.

cheting effect serves as a form of "selection pressure" to order the composition of the elemental parts (see also Wicken, 1984).

Of course, such selection pressure will not be apparent over short intervals, because feedback in biological systems is rarely, if ever, instantaneous. Therefore, in order for an event to be perceived as its own cause, it is first necessary for the effect to complete at least one trip around the feedback loop. It follows, then, that autonomous behavior and its associated selection pressure can be manifest only over intervals exceeding the circuit time of the cycle in question. In a system with multiple feedback loops, the minimal observation time should exceed the interval necessary to complete the slowest feedback route.

By this point, the reader may have noted two very strong objections to the foregoing argument. First of all, the ideal causal loop of Figure 4.1 is an abstraction. No causal structure has been observed to exist without exogenous cause (inputs); no transfer can be effected without losses. In terms of real flows, a cycle appears more like the one depicted in Figure 4.3

But to say that isolated ideal cycles do not exist is not to say that they cannot be regarded as being embedded in real structures. For example, the network in Figure 4.3 may be decomposed into a purely nonautonomous system and an ideal cycle as in Figure 4.4.

True, this separation into cycle and chain is a conceptual artifact. However, it should be stressed that the composite network is influenced by the behavior of *both* component structures. That is, it is insufficient to regard the cycle in Figure 4.4b as merely analogous to a waterwheel being rotated by the flow past it (the flow through the chain in Figure 4.4a). To turn the analogy around, it is just as conceivable that the turning of the waterwheel is assisting the flow past it. In other words, the flows in the nonautonomous chain (Figure 4.4a) might occur at much reduced levels, or might not exist at all, were that chain not coupled with the hypothesized positive feedback inherent in Figure 4.4b. Although ideal loops

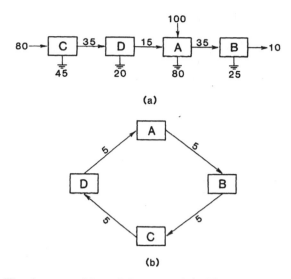

FIGURE 4.4. The decomposition of the network in Figure 4.3 into a straight chain (a) and a closed cycle (b).

cannot exist alone, the properties associated with them do not disappear simply because they are embedded in a larger, less autonomous structure.

The second concern about ideal cycles is the possibility of unlimited positive feedback. In an ideal cycle, positive feedback will probably be destabilizing because an increase anywhere in the loop will be reflected upon itself without attenuation, and eventually the upward spiral will exceed any conceivable bounds. (The presence of any positive feedback in purely linear systems is known to lead to instability. During the early 1970s, when linear analysis of ecosystem stability was popular, the idea of positive feedback in ecosystems was definitely out of favor.) However, it has just been pointed out that ideal cycles cannot exist in isolation; they are always embedded into dissipative networks.

Now, dissipation usually acts to attenuate a signal (cause, flow) as it traverses a cycle. In fact, ecosystems are acknowledged to be highly dissipative structures. Trophic efficiencies of energy or carbon transfer may at times be 10% or less. Although positive feedback is likely to be pathological when it occurs in systems with high efficiencies, in systems with high dissipation, positive feedback may remain completely within bounds. In fact, the argument could be advanced that under heavily dissipative circumstances, positive feedback is the *only mechanism vigorous enough* to maintain structure in the face of such losses. Under less severe conditions, positive feedback is likely to be pathological and undesirable; but under far-from-equilibrium (i.e., highly dissipative) circumstances, it can become a facilitator, if not the principal generator, of system structure.

Also, the notion of positive feedback as a destabilizing factor stems from analyses where the characteristic circuit times are small. If there are appreciable time lags in the circuit, the actual effect of the positive feedback will depend upon the phasing of the signal as it returns to its origin. If the phasing is right, the positive feedback could actually help to stabilize the system (Nisbet, personal communication).

Example 4.1

Positive feedback may serve not only to change the characteristics of its constituents, but also to select potential replacements for its parts. In Figure 4.5a, a new migrant or mutant species, E, has been added to the

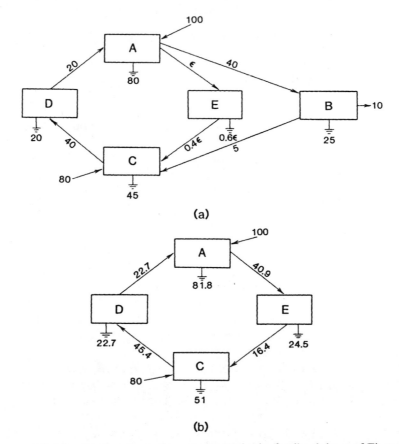

(a)

(b)

FIGURE 4.5. The displacement of component B in the feedback loop of Figure 4.3 by a more efficient species E. (a) The new species enters the network. (b) E totally displaces B.

ecosystem cycle depicted in Figure 4.3. E is slightly more efficient than B at passing medium from A to C and loses no medium to export. At first, E might draw only a miniscule amount, ε, from A. However, even if nothing else in the system changes (a conservative assumption), there will always be a slight reward for A to route more of its output to E and away from B. Eventually, E could displace B from the cycle (Figure 4.5b).

Of course it could always be argued that E won out over B because of proximal events involving A, E, and B. However, looking on matters at the scale of the entire system, it becomes evident that E's function in the entire feedback loop was the chief reason for its success. The feedback loop is seen to survive in Figure 4.5b, albeit with a slightly altered composition. Component B, however, has passed from the scene. One is not being unduly credulous by imagining that A, C, and D could similarly be replaced by other species, giving rise to a situation where the loop endures after all of its original constituent parts have ceased to exist.

Now, it is possible to conceive of the original members of the loop continuing to influence changes in the cycle even after they have disappeared. However, it is highly unlikely that they will fully *determine* the subsequent "behavior." To the degree that the later actions of the cycle are decoupled from the antecedent composition, the loop may be said to behave in an autonomous fashion. What is important here is that this autonomy is not simply the result of impinging exogenous factors. The overall configuration of the system has intervened to select from among the (possibly) chance influences those that are incorporated into its later behavior.

The cycle in Figure 4.5b is highly dissipative. The 11.4 flow units of incremental production generated by E are attenuated to 2.7 units of additional input to A.

Once a positive feedback cycle has been engaged, it takes on a degree of autonomous existence. Barring any perturbation strong enough to obliterate the cycle, components of the structure may come and go like actors in a play, while the play itself (the loop) retains its identity. Positive feedback cycles appear to be fundamental to autonomous behavior.

4.2 Autonomous Behavior and Holistic Description

The plausibility of autonomous behavior by cycles gives new vitality to the call for "holistic" descriptions of systems. Here holism is taken to mean a description of the system-level properties of an ensemble, rather than simply an exhaustive description of all the components (as happens with the reductionistic models referred to earlier). It is thought that by adopting a holistic viewpoint, certain properties become apparent and other behaviors are made visible that otherwise would go undetected.

Such assertions have an air of the metaphysical about them, and it is not surprising that many conscientious scientists are skeptical, if not scornful, of holistic approaches.

However, it has already been pointed out in Chapter 2 that one may pass from the realm of molecules to that of everyday experience and suddenly see properties (pressure, temperature, etc.) that are obscure, to say the least, with microscopic vision. Might it also be that by expanding the field of observation, one sees a greater number of complete causal cycles, so that autonomous behavior appears to emerge as a property of the larger system? Suppose for example, that positive feedback loops exist at all scales of phenomena. One begins with a field of vision limited to a certain small portion of the universe. One then catalogues the behavior of the delimited system, as well as all those influences that cross the system boundary in any direction. If it happens that the original system contains only a segment of a feedback loop, then the behavior of this causal pathway will appear strictly nonautonomous, as in Figure 4.2. However, if one enlarges the system boundary so as to include the entire feedback loop (Figure 4.6), the behavior of the entire loop appears as something that is not entirely dependent upon exogenous causes (i.e., is at least semiautonomous). The semi-autonomous behavior "emerges" from an increase in scope. Continual increase in the extent of the system boundary will enclose further loops, revealing additional autonomous behaviors. It will also diffuse the causality of external influences and heighten the importance of indirect effects (Patten, personal communication).

How one chooses to delimit a system is at the discretion of the observer. If causal loops are evenly distributed at all scales of the universe, it would make little difference how the boundaries are defined. If, however, the loops tend to cluster at certain scales (as experience seems to indicate), then it makes sense to define concentric system boundaries so that each demarcation intersects as few causal loops as possible. One

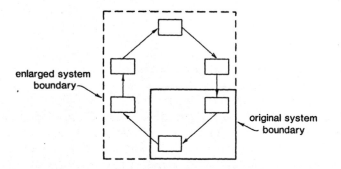

FIGURE 4.6. Enlarging the scope of a system to include an entire feedback loop and its concomitant causal cycle, which may behave in an autonomous manner.

thereby defines a hierarchy of systems (Webster, 1979); for example, cells, organs, organisms, populations, ecosystems, etc. By drawing the boundary around the surface of an organism, for instance, most of the control loops of the animal are internalized. One has to do a great deal less bookkeeping than if one drew the boundary through the midsection of the creature or between the hemispheres of the brain. More importantly, the whole animal is usually more autonomous than only the rear half of the organism!

In light of this apologia for autonomous behavior, certain attitudes still popular in biology appear unduly restrictive, and possibly counter-productive. For example, the neo-Darwinian synthesis maintains that it is nonsensical to speak of the evolution or development of anything beyond the scale of a population. Others will concede the possibility of two populations coevolving, but rare indeed are those individuals who will publicly admit that it might be possible for entire ecosystems to undergo anything akin to development. Therefore, the prevailing view is that the system boundaries must be drawn strictly around each population. And yet these very same populations engage in mass and energy transfers with each other in networks that demonstrably include directed cycles—cycles that are necessarily fragmented by restricting the scope of system boundaries (Patten and Odum, 1981).

One can only wonder what fuels such a proscriptive attitude. Certainly it is not the defense of Darwinian theory. Darwinists are always speaking of fitness for the environment. The biogeochemical cycles in which every living being participates are most assuredly a part of any creature's environment. If it were possible to quantify the autonomous attributes of communities, then one's understanding of fitness would markedly improve.

4.3 The Amount of Cycling in Flow Networks*

More could be said in a qualitative vein about the importance of cycles in describing systems behavior, but eventually the discussion must take a quantitative turn. In Section 3.3, it was shown how the network of direct flows could be analyzed in detail. Input-output analysis can also be used to estimate that fraction of the total throughput being cycled (Finn, 1976, 1980; Patten et al., 1976).

The matrix $[F]$ was derived from the matrix of direct transfers $[T]$ by normalizing each individual flow by the total output of the species from which it originates. In other words, the i-jth element of $[F]$ represents the

---

* The footnote at the beginning of Section 3.3 applies as well to Sections 4.3 and 4.4.

fraction of the throughput of i, which flows directly to j. The reader will recall that the i-jth element of $[F]^m$ represents the fraction of the throughput of i that flows to j along all possible pathways of length m. In particular, each nonzero diagonal element of $[F]^m$ gives the fraction of i's throughput that cycles back to i after exactly m transfers. Now, all the integer powers of the $[F]$ matrix (including the zeroeth power, or identity matrix) sum to give the output structure matrix as in (3.11). The right-hand side of (3.11) makes it clear that the diagonal components of the output structure matrix are at least 1, and any diagonal entry exceeding unity implies that the designated compartment engages in cycling.

Example 4.2

The matrix of host coefficients $[F]$ for energy flow in Cone Spring (depicted in Figure 3.2) is:

$$\begin{bmatrix} 0 & .794 & 0 & 0 & 0 \\ 0 & 0 & .453 & .201 & 0 \\ 0 & .307 & 0 & .014 & 0 \\ 0 & .084 & 0 & 0 & .155 \\ 0 & .451 & 0 & 0 & 0 \end{bmatrix},$$

so that the output structure matrix $[I - F]^{-1}$ becomes:

$$\begin{bmatrix} 1 & .958 & .434 & .199 & .031 \\ 0 & 1.207 & .547 & .251 & .039 \\ 0 & .374 & 1.169 & .092 & .014 \\ 0 & .186 & .084 & 1.039 & .161 \\ 0 & .545 & .247 & .113 & 1.018 \end{bmatrix}.$$

Thus, 100 units leaving the bacteria (component 3) at a given instant eventually result in 116.9 units of flow through the bacteria. In addition to the original 100 units, 16.9 eventually flow back through the bacteria after having traversed one or more cycles.

Not surprisingly, the diagonal elements of the output structure matrix $S(= [I - F]^{-1})$ can be used to calculate that portion of the total system throughput T that is attributable to cycling; for example, T_c. If one unit leaving i causes S_{ii} units of total flow through i, and $(S_{ii} - 1)$ of that throughflow is a result of cycling (see Example 4.2), then the fraction of T_i due to cycling becomes $(S_{ii} - 1)/S_{ii}$. Multiplying each such fraction by its corresponding T_i and summing over all components gives T_c. Finn (1980) appropriately defines a system index of cycling as T_c/T—a fraction that may vary from 0 (no cycling) to 1 (complete cycling).

Example 4.3

To calculate the cycling index for Cone Spring, one employs the diagonal elements of the structure matrix as calculated in Example 4.2, along with the compartmental throughflows that were determined in Example 3.5, and substitutes these values into the formula

$$T_c = \sum_i T_i(S_{ii} - 1)/S_{ii}$$

$$= 11,184(1 - 1)/1 + 11,483(1.207 - 1)/1.207 + 5205(1.169 - 1)/1.169$$

$$+ 2384(1.039 - 1)/1.039 + 370(1.018 - 1)/1.018$$

$$= 0 + 1967.0 + 753.8 + 88.5 + 6.4$$

$$= 2815.7 \text{ kcal m}^{-2} \text{ y}^{-1}.$$

The cycling index is 2816/42,445, or 6.63%. (This value is less than the 9.2% calculated by Finn [1980], because Finn did not include all the systems flows in his definition of total system throughput. The calculation of T_c is unaffected by this difference in terminology.)

Odum (1969) identified a greater degree of cycling as an attribute of more mature ecosystem networks. To test this hypothesis, Richey et al. (1978) looked at the cycling index of carbon flow in five freshwater lakes with varying degrees of eutrophication. Unfortunately, the results were inconclusive, indicating that the gross amount of cycling is insufficient to describe the stage of ecosystem development.

4.4 The Structure of Network Cycles

The cycling index is a true macroscopic (community-level) property of a flow network. Its weak correlation with the common notion of ecosystem development forces a continued search for a more suitable macroscopic indicator of growth and maturation. However, having already spent most of this chapter discussing cycles and feedback as microscopic agents of autonomous development, it should not inordinately detract from the flow of this book to spend the remainder of the present chapter showing how one might delineate the microscopic structure of cycles in flow networks.

If one wishes to investigate the structure of cycling in networks, probably the first question that comes to mind is: "Which are the major pathways that recycle media?" To treat this matter exhaustively, one must know the entire suite of potential pathways for recycle; that is, possess an enumerated list of all simple direct cycles in the network (i.e., cycles with no repeated elements). Next, this list must be ordered in some fashion. More flow is likely to recycle along some circuits than over others. Is it possible to assign to each loop a relative amount of flow that characterizes that circuit's importance as a pathway for recycle? Within each cycle, which link is most critical to the established rate of recycle?

Finally, because it was argued in Section 4.1 that feedback loops may be primary factors influencing network structure, it becomes potentially useful to know how the cycled flow is distributed over the entire network of flows. To achieve this last goal, it is useful to imagine separating the cyclic from the once-through flow; that is, dividing the actual network into two virtual networks: (1) the superposition of all the individual closed cycles; and (2) a tree-like structure of unidirectional flows. The interdependence of these two component networks has already been discussed briefly in connection with Figure 4.3.

All of these issues are addressed below in quantitative terms; albeit, neither in the order just presented, nor in a totally unequivocal way.

Example 4.4

From the network diagram of energy flow in Cone Spring (Figure 3.2), it is possible to identify by inspection the five distinct simple cycles indicated in Figure 4.7. The quantities assigned to each cycle are chosen so that when one subtracts these circuits from the network, no cycles remain among the residual flows. Tracing around the last cycle (detritus-bacteria-detritus), it is clear that one wishes to subtract the smallest arc from the starting network to ensure that the given cycle will be eliminated, but still

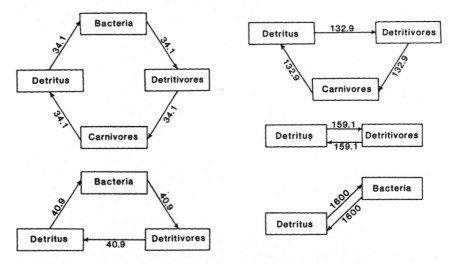

FIGURE 4.7. Five simple directed cycles of energy flow (kcal $m^{-2}y^{-1}$) that may be identified in the Cone Spring network (Figure 3.2). The method for assigning the magnitudes to each cycle is described later in the text. Reprinted by permission of the publisher from Identifying the Structure of Cycling in Ecosystems, by R.E. Ulanowicz, Mathematical Biosciences Vol. 65, copyright 1983 by Elsevier Science Publishing Co., Inc.

guarantee that no arc in the remaining flows will become negative. The situation is complicated in the other cycles by overlapping arcs, but an algorithm valid in the general case is described below.

Subtracting the five cycles from the starting network yields the acyclical graph (tree) in Figure 4.8.

Identifying directed cycles in networks with small numbers of components and arcs is a simple task. Unfortunately, as the number of components grows, enumeration of the cycles can become inordinately difficult. Roughly speaking, the potential number of cycles in a graph increases as the factorial of the number of nodes. When one considers that 20! is about 2×10^{18}, it becomes obvious that the task can exhaust the most ambitious individual and even the fastest computers available.

Fortunately for ecologists, the connectance (the percentage of possible links that are realized) of observed ecosystems networks rarely exceeds 25% for networks with many species. Nonetheless, the search may still be tedious, and it behooves the investigator to choose an efficient algorithm to identify the cycles. Mateti and Deo (1976) have reviewed the methods of enumerating cycles in graphs and have concluded that backtracking search algorithms with suitable pruning methods (to eliminate searching many futile pathways) are the most efficient programs in the greatest number of instances.

In backtracking algorithms, one orders the nodes in some convenient way (described below) and imagines the same order of n nodes to be

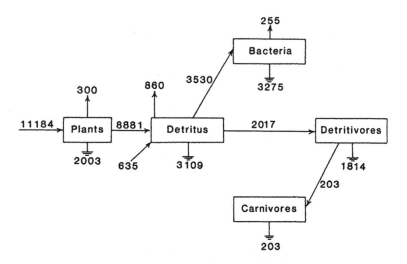

FIGURE 4.8. The residual acyclic network of flows (kcal m^{-2} y^{-1}) obtained after the five cycles depicted in Figure 4.7 have been subtracted from the original Cone Spring network (Figure 3.2). Reprinted by permission of the publisher from Identifying the Structure of Cycling in Ecosystems, by R.E. Ulanowicz, Mathematical Biosciences, Vol. 65. Copyright 1983 by Elsevier Science Publishing Co., Inc.

	n	[1]	[2]	[3]	. . .	[n]
l	.	.	.	.	. . .	.
e	:	:	:	:	. . .	:
v	2	[1]	[2]	[3]	. . .	[n]
e	1	[1]	[2]	[3]	. . .	[n]
l	0	[1]	[2]	[3]	. . .	[n]

compartment

FIGURE 4.9. Mnemonic diagram useful in keeping track of the operations in the backtracking algorithm written to identify directed cycles in an n-component network.

repeated at n levels as in Figure 4.9. One begins at a given node in the zeroeth level (designated as the pivot element) and searches from left to right among nodes in the next level until an existing flow connection is found. One then jumps to the chosen node in the next level and proceeds searching (left to right) the level above it in an attempt to move higher. As one progresses to higher levels, the last node visited in each previous level is stored in a vector array which describes a current pathway. Before advancing to a higher level one checks to make sure that the new node has not already appeared in the current pathway. One ascends to as high a level as possible until interrupted by one of two circumstances: (1) If an arc exists to the pivot element in the next level, then a simple directed cycle has been identified. Its description is read from the current pathway, and the search continues. (2) If one is searching from node k in level m and all possibilities in level $m + 1$ have been exhausted (i.e., one can move no further to the right), then one backtracks (whence the name) to the node in the current pathway at the $m - 1$th level and begins searching the mth level starting with the $k + 1$th node. When further backtracking becomes impossible, all cycles passing through the pivot element have been identified. The pivot element may be eliminated from further consideration, thereby decreasing the dimension of the subsequent search.

Example 4.5

In order to apply the backtracking algorithm to the Cone Spring network, it helps to consider the compartments in the order 2, 3, 4, 5, 1. One uses either Figure 3.2 or the matrix $[F]$ in Example 4.2 to test for actual connections. The following mnemonic array below may help to keep track of the order of operations:

l	4	2	3	4	5	1
e	3	2	3	4	5	1
v	2	2	3	4	5	1
e	1	2	3	4	5	1
l	0	2	3	4	5	1

compartment

Beginning with pivot element 2 in level 0, one searches level 1 from left to right following the instructions in the text above. The order of the significant operation in the search and the current pathways they generate are as follows:

Operation	Current pathway
Begin at pivot element	2
Advance to level 1	2-3
Report cycle 1	2-3-2
Advance to level 2	2-3-4
Report cycle 2	2-3-4-2
Advance to level 3	2-3-4-5
Report cycle 3	2-3-4-5-2
Backtrack to level 2	2-3-4
Backtrack to level 1	2-3
Backtrack to level 0	2
Advance to level 1	2-4
Report cycle 4	2-4-2
Advance to level 2	2-4-5
Report cycle 5	2-4-5-2
Backtrack to level 1	2-4
Backtrack to level 0	2
Further backtracking impossible. END	—

It accidentally happens that all cycles contain node 2, so that searches starting from the remaining four pivot elements uncover no further cycles.

If the first pivot element in Example 4.5 had been component 1, significant time would have been wasted searching for cycles containing the plants. (There are none.) It is obvious that the order of the pivot elements strongly influences the time it takes to complete the full search.

For the algorithm to be most efficient, it is necessary to first treat those compartments that are most likely to complete a cycle. Now, starting at any given node in the network, it is a relatively easy task to identify all the other compartments that may be reached along a simple pathway from the "root" or starting node (Knuth, 1973). Any arc from a "reachable" node back to the root node is called a cycle arc; because, as its name implies, it completes a cycle. Hence, one may queue the pivot elements in decreasing order of the number of associated cycle arcs. (Those nodes with no associated cycle arcs, like node 1 in Example 4.4, may be excluded from the backtracking algorithm.) Although counting the cycle arcs does take time (roughly proportional to the cube of the number of nodes), it is more than recouped by decreasing the time spent searching networks of even modest complexity. The existing, quantified ecological flow networks rarely exceed 25 compartments. As long as their connectivity remains modest, ordering the pivot elements by number of cycle arcs is usually sufficient to keep the search time within reasonable bounds. Those interested in further "pruning" methods are referred to Read and Tarjan (1975).

Enumerating the simple cycles in a network is only the first step in fully describing the structure of cycling in a network. Typically, the number of simple cycles runs into the hundreds, and it becomes necessary to systematically aggregate the cycles into a more tractable number of groupings. Because the number of cycles usually exceeds the number of arcs in the network, significant overlap among the many cycles is bound to occur. When that overlap includes crucial transfers, the coinciding arcs may be used to aggregate the cycles.

The critical or most vulnerable arc in a cycle bears analogy to the weakest link in a chain (a cycle being a chain turned back upon itself). How one identifies the critical arc in a cycle is subject to individual interpretation—the physiologist might hold opinions different from those held by an ethologist. Rather than belabor the possibilities, an analogy will be made with chemical kinetics, where a sequence of rates is considered to be controlled by the slowest step. Accordingly, the assumption is made here that the critical arc in a cycle is the one possessing the smallest (slowest) flow. Other interpretations are, of course, possible. It then becomes an easy task to identify the critical arc in each simple cycle, and all cycles sharing a common critical arc will be dominated by flow along that link. The set of critical arcs, therefore, defines an equal number of cycle groupings, or "nexuses."

In the last chapter, it was argued that the effects of myriad phenomena were implicit in the measured flows. At the end of Section 4.1, it was further suggested that the feedback phenomenon associated with the loops was a key determinant of structure in dissipative networks. This suggests the hypothesis that the set of critical arcs pinpoints the controlling transfers in an ecosystem. They are likely to be the loci of either strain or growth, and any changes in a critical arc should be propagated most strongly over the nexus it defines.

Example 4.6

One of the best sets of data on ecological flows describes the flux of carbon through a tidal marsh creek ecosystem adjacent to Crystal River,

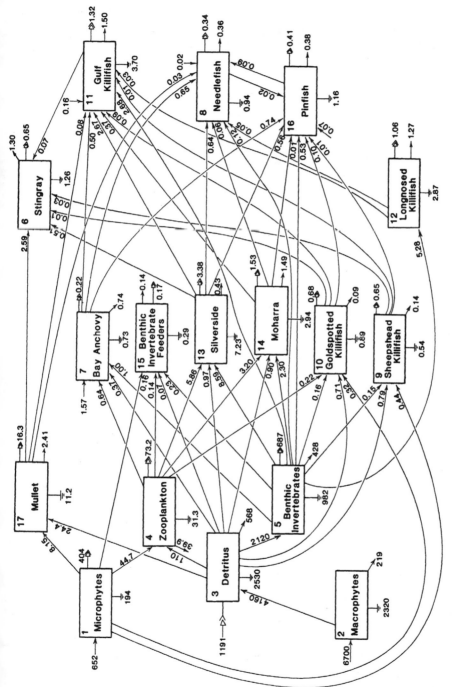

FIGURE 4.10. A schematic of carbon flows (mg m⁻² d⁻¹) among the taxa of a marsh gut ecosystem, Crystal River, Florida. The linked arrows depict returns to the detritus (after M. Homer and W. M. Kemp, unpublished manuscript). Reprinted by permission of the publisher from Identifying the Structure of Cycling in Ecosystems, by R. E. Ulanowicz, Mathematical Biosciences, Vol. 65. Copyright 1983 by Elsevier Science Publishing Co., Inc.

Florida (Homer and Kemp, unpublished ms.; see also Ulanowicz, 1983). The 17 components of the ecosystem consist mostly of vertebrate fishes. The flows of carbon in mg m^{-2} d^{-1} are depicted as arrows in Figure 4.10. The network consists of 6 exogenous inputs, 16 exports, 17 respiratory flows, and 69 internal exchanges.

A total of 119 simple cycles are embedded in the network as listed in Table 4.1. The cycles segregate into 41 nexuses, several of which contain numerous cycles. For example, the predation on Gulf killifish by stingray dominates a 14 cycle nexus, pinfish feeding on needlefish defines a 13 cycle nexus, the reciprocal flow from pinfish to needlefish influences a separate 10 cycle nexus, etc.

How would a change in a critical arc be reflected in this analysis of cycles? It happens that in those networks that have been analyzed by this methodology, the larger (in number of cycles) and more complex (constituent cycles are longer) nexuses were defined by smaller flows occurring at higher trophic levels. Because these higher trophic transfers are contingent upon many events at lower levels and because perturbations occuring anywhere over the widespread nexus will focus upon the critical arc, one expects the larger nexuses to be early victims of stress on the system.

Example 4.7

The comparative study of ecosystems networks suffers from a severe lack of data on comparable networks that have been parsed using identical criteria. A notable exception is the Crystal River study cited in the previous example. Homer and Kemp sampled a nearby tidal creek identical to the one in Example 4.6 in all aspects, except that the second creek was subjected to a continuous 6°C elevation in temperature because of heated effluent from a nearby nuclear power generating station. The network of the perturbed system is depicted in Figure 4.11.

More drastic than the decrease in productivity and total system throughput is the altered structure of cycling in the network. Table 4.2 lists the 30 nexuses of 46 total cycles present in the perturbed network. All of the "larger" nexuses identified in Table 4.1 have disappeared from the heated creek system. The largest nexus in the disturbed system is a single, 4 cycle nexus defined by the losses from pinfish to the detritus.

Finn's cycling index actually rose from 7.1% in the control creek to 9.4% in the stressed system—presumably a homeostatic response to disturbance. In fact, the overall picture of change resembles eutrophication: The higher order (presumably slower) nexuses have disappeared. The shorter, faster, trophically lower cycles survive and turn over more intensely.

TABLE 4.1. Enumeration of the 119 cycles
embedded in Figure 4.10 aggregated
according to critical arc

3 cycle nexus with critical arc (10, 6) = .010
1. 3-10-6-3
2. 3-4-10-6-3
3. 3-5-10-6-3
6 cycle nexus with critical arc (14, 16) = .010
4. 3-14-16-3
5. 3-14-16-8-3
6. 3-4-14-16-3
7. 3-4-14-16-8-3
8. 3-5-14-16-3
9. 3-5-14-16-8-3
4 cycle nexus with critical arc (9, 11) = .010
10. 3-9-11-3
11. 3-9-11-6-3
12. 3-5-9-11-3
13. 3-5-9-11-6-3
4 cycle nexus with critical arc (9, 16) = .010
14. 3-9-16-3
15. 3-9-16-8-3
16. 3-5-9-16-3
17. 3-5-9-16-8-3
13 cycle nexus with critical arc (8, 16) = .020
18. 3-7-8-16-3
19. 3-13-8-16-3
20. 3-14-8-16-3
21. 3-4-7-8-16-3
22. 3-4-13-8-16-3
23. 3-4-14-8-16-3
24. 3-5-8-16-3
25. 3-5-7-8-16-3
26. 3-5-13-8-16-3
27. 3-5-14-8-16-3
28. 3-5-12-8-16-3
29. 3-17-8-16-3
30. 8-16-8
2 cycle nexus with critical arc (9, 6) = .030
31. 3-9-6-3
32. 3-5-9-6-3
2 cycle nexus with critical arc (12, 11) = .030
33. 3-5-12-11-3
34. 3-5-12-11-6-3
1 cycle nexus with critical arc (17, 8) = .030
35. 3-17-8-3
1 cycle nexus with critical arc (12, 8) = .050
36. 3-5-12-8-3
3 cycle nexus with critical arc (14, 8) = .060
37. 3-14-8-3
38. 3-4-14-8-3
39. 3-5-14-8-3
14 cycle nexus with critical arc (11, 6) = .070

TABLE 4.1. *Continued*

40. 3-7-11-6-3
41. 3-10-11-6-3
42. 3-13-11-6-3
43. 3-14-11-6-3
44. 3-4-7-11-6-3
45. 3-4-10-11-6-3
46. 3-4-13-11-6-3
47. 3-4-14-11-6-3
48. 3-5-11-6-3
49. 3-5-7-11-6-3
50. 3-5-10-11-6-3
51. 3-5-13-11-6-3
52. 3-5-14-11-6-3
53. 3-17-11-6-3
    1 cycle nexus with critical arc (3, 15) = .070
54. 3-15-3
    1 cycle nexus with critical arc (17, 11) = .080
55. 3-17-11-3
    10 cycle nexus with critical arc (16, 8) = .090
56. 3-7-16-8-3
57. 3-10-16-8-3
58. 3-13-16-8-3
59. 3-4-7-16-8-3
60. 3-4-10-16-8-3
61. 3-4-13-16-8-3
62. 3-5-16-8-3
63. 3-5-7-16-8-3
64. 3-5-10-16-8-3
65. 3-5-13-16-8-3
    3 cycle nexus with critical arc (10, 11) = .090
66. 3-10-11-3
67. 3-4-10-11-3
68. 3-5-10-11-3
    3 cycle nexus with critical arc (10, 16) = .100
69. 3-10-16-3
70. 3-4-10-16-3
71. 3-5-10-16-3
    1 cycle nexus with critical arc (5, 8) = .120
72. 3-5-8-3
    1 cycle nexus with critical arc (4, 15) = .140
73. 3-4-15-3
    1 cycle nexus with critical arc (5, 9) = .150
74. 3-5-9-3
    1 cycle nexus with critical arc (15, 3) = .170
75. 3-5-15-3
    3 cycle nexus with critical arc (7, 3) = .220
76. 3-7-3
77. 3-4-7-3
78. 3-5-7-3
    1 cycle nexus with critical arc (4, 10) = .220
79. 3-4-10-3
    6 cycle nexus with critical arc (8, 3) = .340

TABLE 4.1. *Continued*

80. 3-7-8-3
81. 3-13-8-3
82. 3-4-7-8-3
83. 3-4-13-8-3
84. 3-5-7-8-3
85. 3-5-13-8-3
    2 cycle nexus with critical arc (3, 7) = .370
86. 3-7-11-3
87. 3-7-16-3
    3 cycle nexus with critical arc (14, 11) = .370
88. 3-14-11-3
89. 3-4-14-11-3
90. 3-5-14-11-3
    6 cycle nexus with critical arc (16, 3) = .410
91. 3-13-16-3
92. 3-4-7-16-3
93. 3-4-13-16-3
94. 3-5-16-3
95. 3-5-7-16-3
96. 3-5-13-16-3
    2 cycle nexus with critical arc (7, 11) = .500
97. 3-4-7-11-3
98. 3-5-7-11-3
    3 cycle nexus with critical arc (13, 6) = .510
99. 3-13-6-3
100. 3-4-13-6-3
101. 3-5-13-6-3
    1 cycle nexus with critical arc (5, 10) = .610
102. 3-5-10-3
    1 cycle nexus with critical arc (9, 3) = .650
103. 3-9-3
    1 cycle nexus with critical arc (6, 3) = .650
104. 3-17-6-3
    1 cycle nexus with critical arc (10, 3) = .680
105. 3-10-3
    1 cycle nexus with critical arc (3, 14) = .900
106. 3-14-3
    2 cycle nexus with critical arc (3, 13) = .970
107. 3-13-3
108. 3-13-11-3
    1 cycle nexus with critical arc (12, 3) = 1.060
109. 3-5-12-3
    3 cycle nexus with critical arc (11, 3) = 1.320
110. 3-4-13-11-3
111. 3-5-11-3
112. 3-5-13-11-3
    2 cycle nexus with critical arc (14, 3) = 1.530
113. 3-4-14-3
114. 3-5-14-3
    2 cycle nexus with critical arc (13, 3) = 3.380
115. 3-4-13-3

TABLE 4.1. *Continued*

116. 3-5-13-3
     1 cycle nexus with critical arc (17, 3) = 16.290
117. 3-17-3
     1 cycle nexus with critical arc (4, 3) = 73.200
118. 3-4-3
     1 cycle nexus with critical arc (5, 3) = 686.900
119. 3-5-3

The choice of the critical arc as the smallest link in a cycle affords a mathematical convenience. As mentioned in Example 4.4, one may eliminate a cycle from the network by subtracting the magnitude of the weakest arc from each arc in the cycle. Of course, the residual arcs remain nonnegative. From the definition of a nexus, it follows that the removal of the critical arc severs all the cycles in the nexus. The only question remaining is how to distribute the magnitude of the critical flow among all the cycles of the nexus. Any of a number of choices is possible. For example, the flow could be divided equally among the cycles of the nexus, or it could be allotted to each cycle in proportion to the contribution of that cycle to some specified community property, such as total flow. Perhaps the most intuitively pleasing scheme (and the one used here) is to prorate the critical flow to each cycle according to the probability that a quantum of medium will complete that particular cycle (Silvert, private communication).

Once an apportionment scheme has been chosen, the nexuses may be "peeled off" the network beginning with the smallest critical link. To ensure that no residual flow goes negative, it becomes necessary to redefine the remaining suite of critical links and nexuses after each nexus has been removed. After the last nexus has been removed, the residual network will be acyclical and will retain the original inputs, exports, and dissipations. The network of composite cycled flows will, of course, consist entirely of ideal cycles.

FIGURE 4.11. Schematic of carbon flows (mg m^{-2}d^{-1}) in a marsh gut ecosystem nearby and similar to the one in Figure 4.10, except that it was perturbed by a 6°C average elevation in temperature due to thermal effluent from a nearby power generating station.

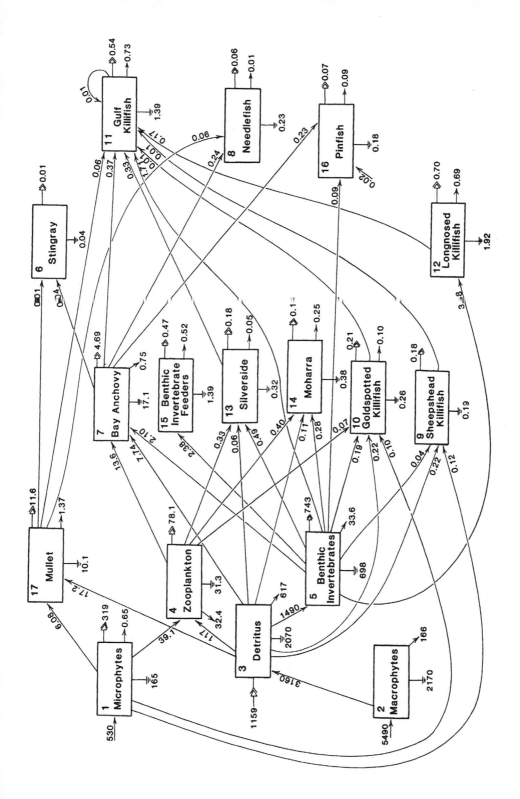

TABLE 4.2. Enumeration of the 46 cycles found in Figure 4.11 aggregated according to critical arc

---

3 cycle nexus with critical arc (6, 3) = .010
1. 3-7-6-3
2. 3-4-7-6-3
3. 3-5-7-6-3
   3 cycle nexus with critical arc (10, 11) = .010
4. 3-10-11-3
5. 3-4-10-11-3
6. 3-5-10-11-3
   2 cycle nexus with critical arc (9, 11) = .010
7. 3-9-11-3
8. 3-5-9-11-3
   1 cycle nexus with critical arc (17, 6) = .010
9. 3-5-17-6-3
   1 cycle nexus with critical arc (11, 11) = .010
10. 11-11
    1 cycle nexus with critical arc (5, 9) = .040
11. 3-5-9-3
    3 cycle nexus with critical arc (8, 3) = .060
12. 3-7-8-3
13. 3-4-7-8-3
14. 3-5-7-8-3
    2 cycle nexus with critical arc (3, 13) = .060
15. 3-13-3
16. 3-13-11-3
    1 cycle nexus with critical arc (17, 11) = .060
17. 3-5-17-11-3
    1 cycle nexus with critical arc (17, 8) = .060
18. 3-5-17-8-3
    4 cycle nexus with critical arc (16, 3) = .070
19. 3-7-16-3
20. 3-4-7-16-3
21. 3-5-7-16-3
22. 3-5-16-3
    1 cycle nexus with critical arc (4, 10) = .070
23. 3-4-10-3
    1 cycle nexus with critical arc (3, 14) = .110
24. 3-14-3
    2 cycle nexus with critical arc (14, 3) = .160
25. 3-4-14-3
26. 3-5-14-3
    1 cycle nexus with critical arc (12, 11) = .170
27. 3-5-12-11-3
    1 cycle nexus with critical arc (9, 3) = .180
28. 3-9-3
    2 cycle nexus with critical arc (13, 3) = .180
29. 3-4-13-3
30. 3-5-13-3
    1 cycle nexus with critical arc (5, 10) = .190
31. 3-5-10-3
    1 cycle nexus with critical arc (10, 3) = .210

TABLE 4.2. *Continued*

32. 3-10-3
    1 cycle nexus with critical arc (4, 13) = .330
33. 3-4-13-11-3
    1 cycle nexus with critical arc (13, 11) = .330
34. 3-5-13-11-3
    3 cycle nexus with critical arc (7, 11) = .370
35. 3-7-11-3
36. 3-4-7-11-3
37. 3-5-7-11-3
    1 cycle nexus with critical arc (15, 3) = .470
38. 3-5-15-3
    1 cycle nexus with critical arc (11, 3) = .540
39. 3-5-11-3
    1 cycle nexus with critical arc (12, 3) = .700
40. 3-5-12-3
    1 cycle nexus with critical arc (5, 7) = 2.100
41. 3-5-7-3
    2 cycle nexus with critical arc (7, 3) = 4.690
42. 3-7-3
43. 3-4-7-3
    1 cycle nexus with critical arc (17, 3) = 11.640
44. 3-5-17-3
    1 cycle nexus with critical arc (4, 3) = 78.100
45. 3-4-3
    1 cycle nexus with critical arc (5, 3) = 742.600
46. 3-5-3

Example 4.8

The Crystal River control creek depicted in Figure 4.10 may be decomposed into purely cycled flow (Figure 4.12) and residual acyclical flow (Figure 4.13).

Since an acyclical network can be found within any given food web, it becomes possible to aggregate this underlying network into a concatenated trophic chain of n or fewer levels, as outlined in Section 3.3.

Having argued that cycles are a primary influence upon network organization, it is now useful to shift attention to certain mathematical tools that are useful in quantifying that organization.

4.5 Summary

It is difficult to unambiguously define causes in networks with cycles. In describing circular causality, it is hard to avoid what appears to be circular reasoning. Hence, most prefer to avoid discussing circular causality and to frame description in reductionistic terms. However, this is a hazardous alternative in that the reductionistic perspective unavoidably ex-

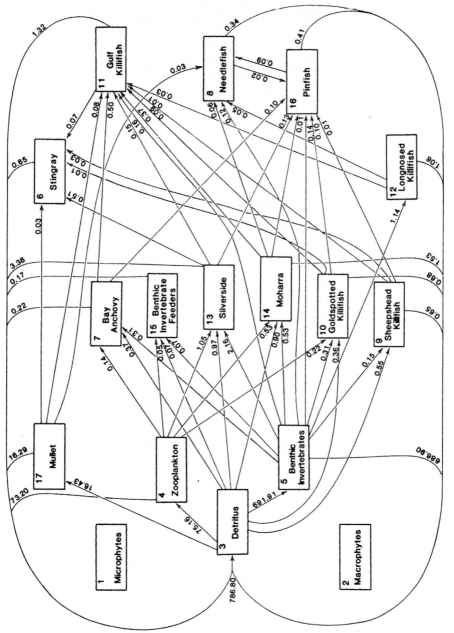

FIGURE 4.12. The composite nexus of all cycled flow inherent in the network in Figure 4.10. See Figure 4.10 legend for further details. Reprinted by permission of the publisher from Identifying the Structure of Cycling in Ecosystems, by R. E. Ulanowicz, Mathematical Biosciences, Vol. 65. Copyright 1983 by Elsevier Science Publishing Co., Inc.

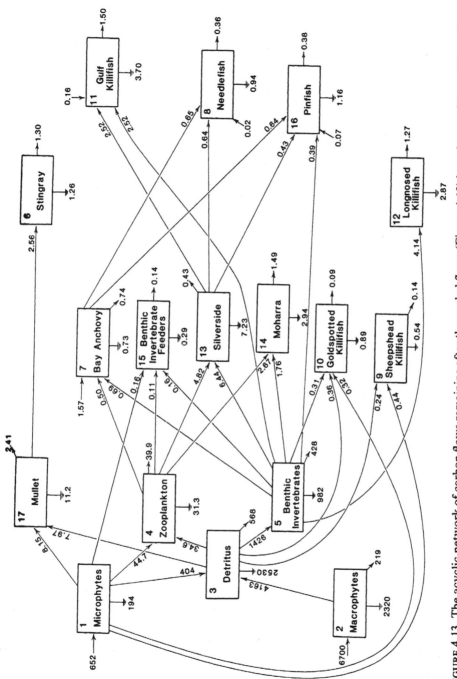

FIGURE 4.13. The acyclic network of carbon flows remaining after the cycled flows (Figure 4.12) have been subtracted from the original flows (Figure 4.10). Reprinted by permission of the publisher from Identifying the Structure of Cycling in Ecosystems, by R. E. Ulanowicz, Mathematical Biosciences, Vol. 65. Copyright 1983 by Elsevier Science Publishing Co., Inc.

cludes discussion of the autonomous behavior of feedback loops in the system.

The crux of autonomous behavior lies in the tendency of simple direct cycles to exhibit positive feedback. Positive feedback can act to select from among variations occurring within the cycle. The behavior of the cyclical structure can strongly influence the makeup of its constituents. Causality can thereby flow down the hierarchy, but this influence can be observed only at spatial scales that encompass whole cycles and over temporal intervals that exceed the period of the cycle. It is fashionable to proscribe discussion of living phenomena at these larger temporal and spatial scales; but this attitude may be overly restrictive and possibly counter productive.

The amount of flow cycling in a network can be quantified using input-output analysis. However, it is sometimes necessary to know more about the actual structure of cycling in the network. Systematic methods exist to identify all the simple cycles in a network. The elemental cycles may be grouped according to shared critical arcs. This same aggregation procedure helps to hypothetically separate the cycled flow from the rest of the system.

5
The Calculus

"The kinds of questions we ask are as many as the kinds of things which we know."

Aristotle
Analytica Posteriora

5.1 Information Theory and Ecology

In the last two chapters, flows have been discussed as if they were continuous processes. However, biological transformations, especially those in large-scale systems, actually occur as a series of discrete, often unpredictable transfers. The appropriate mathematical language with which to describe such discrete events is probability theory. Probability was the starting point for statistical mechanics, and it remains the basis for a phenomenological description of growth and development.

Now, developing systems are, by definition, changing; and the probabilities of microscopic events internal to maturing systems change accordingly. The study of what gives rise to a change in probability assignment defines the realm of information theory (Tribus and McIrvine, 1971). Information theory as a branch of pure mathematics is seen to be a progeny of probability theory, and is likewise necessary to the description of growth and development.

The mathematical literature on probability and information is legion. Concerning the former, there is little this author could add to the reasonably clear picture most readers have as to what constitutes a probability and how it can be estimated (although those desiring to have their confidence on this topic shaken are referred to Jaynes, 1979). Regarding information theory, however, there appears to be rather widespread confusion (at least in the ecological literature) as to what information is, how it can be measured, and how it may be employed in ecological narrative.

Although elements of information theory appeared in the thermodynamic literature of the 19th Century (Boltzmann, 1872), information theory as a discipline is commonly regarded as having developed from code-breaking activities during World War II and having been given formal, public structure in Shannon's (1948) treatise. Because of the early association with cryptography and communications networks, some individuals still consider information theory to have only a narrow field of application. Fortunately, many others have become aware of how fundamental the theory is, and Brillouin (1956) and Tribus (1961) have gone as far as to reinterpret that most universal of scientific disciplines, thermodynamics, in terms of information theory.

Very soon after information theory became popular, the ecologist MacArthur (1955) demonstrated that information concepts were applicable to ecological flow networks. Specifically, MacArthur invoked the Shannon-Wiener definition of uncertainty to characterize the multiplicity of flow pathways among the populations of an ecosystem. MacArthur's intention was to permit quantitative testing of Odum's (1953) earlier hypothesis that more parallel pathways between any two compartments of an ecosystem afforded greater homeostatic regulation of flows.

Odum's idea was quite simple. If a population depended upon a single concatenation of transfers to supply a necessary element, that supply was highly vulnerable to perturbation at any point along the line. If, however, supplier and recipient were linked by several parallel pathways, it became more likely that a perturbation of flow along any given pathway could be compensated by opposite changes in flows along the less impacted parallel routes (a specific instance of the LeChâtelier-Braun principle). Subjecting ecosystems with different network uncertainties (in the sense of Shannon) to equivalent perturbations should reveal whether homeostasis is, in fact, correlated with flow diversity. Ecology, it seemed, had a community-level (holistic) hypothesis that was testable and apparently well articulated. The number of investigators who became involved with the hypothesis over the next decade was impressive (Woodwell and Smith, 1969).

Unfortunately, the results from possibly hundreds of man-years of work on the diversity-stability hypothesis have been rather inconclusive. Because of this seemingly futile effort, many ecologists now disdain information theory. Anyone seeking to revive interest in the subject is met with overwhelming disinterest. However, information theory is so fundamental to quantitative biology that it is impossible to ignore such a powerful tool. Where, then, did matters go awry with the diversity-stability controversy? The primary reasons appear to be twofold.

First, there was a diversion of attention from flows towards population levels. It is easy to understand such a shift. After all, flows are notoriously difficult and expensive to measure. By contrast, population levels (whether measured in numbers, biomass, or other units) can be quantified by relatively straightforward techniques. Furthermore, the emphasis upon flows by Lindeman, Odum, and MacArthur was somewhat alien to the classical biological concern over the contents of categories. This is not to fault those who first suggested population diversity indices as estimators of system homeostasis—it certainly seemed like a good idea at the time. But it does seem regrettable that it has taken so long to return attention to the interrelationships among flows.

The second difficulty with most early applications of information theory to ecosystems was the rampant confusion about the definition of information. To be more specific, it was common to regard the Shannon-Wiener index (H) as both the amount of uncertainty in a given distribution as well as the amount of information associated with that arrangement. None other than Norbert Wiener (1948) used H in this confusing manner. Thus, communities with high population diversity indices were said to be simultaneously information-rich and chaotic!

Strictly speaking, it is not incorrect to regard H as the information in a given probability distribution, but it is a pedagogical disaster to do so. There are ways to capture the intuitive notion of information in a much clearer and more precise fashion. The unambiguous mathematical definition of information thus becomes the primary objective of this chapter. However, to define information, it is first necessary to consider uncertainty.

5.2 The Uncertainty of an Outcome*

Uncertainty is the inability to say a priori what the exact outcome of a given event will be. A die is thrown; no one can always predict which face will land upward. Fish of a given species are harvested from an estuary; it is impossible to say several years in advance precisely what the yield will be. An organism belonging to a given species will be consumed by one of several possible predators; which predator will it be?

In a very real sense, the uncertainty about the answer to any query depends upon how the question is posed. The answer to a properly posed question always falls into a finite number of discrete categories. These categories must exhaust all possible outcomes of the event. The initial uncertainty depends upon what those categories are. For example, if the throw of a die may fall into six categories corresponding to the integers 1–6, uncertainty about the result takes on a given level. If the question is posed so that the outcome can fall into only two categories (1–3 and 4–6), there is a little less uncertainty about the outcome. If the first category includes the integers 1–5 and the second comprises only the integer 6, matters are still less uncertain. Finally, if the two categories separate the integers 1–6 from the integers 7–10, no uncertainty remains about the outcome.

From this simple example, it may be conjectured that the uncertainty about a particular outcome is related inversely to the probability of that result taking place. In the first question, the equiprobability for each face

---

* Some information theorists may object to assigning an uncertainty to a single category, but the designation is retained here by virtue of its significant pedagogical utility (Higashi, personal communication).

(1/6) determines the level of uncertainty. In the second instance, only two categories are equiprobable (1/2), and the uncertainty about either case is somewhat less. In the asymmetrical third instance, no one is too surprised (reasonably confident) when the first category ($p = 5/6$) is obtained, but they are relatively surprised when a 6 occurs ($p = 1/6$). Finally, there is no surprise whatsoever (full certainty) about the outcome to the fourth question.

The uncertainty about a given outcome should therefore be related to the reciprocal of the probability for that result. The optimal functional form turns out to be logarithmic (see below), so that uncertainty is taken to be proportional to the logarithm of the reciprocal of the probability,

$$H_i = K \log (1/p_i), \tag{5.1}$$

or

$$H_i = -K \log p_i, \tag{5.2}$$

where p_i is the probability of outcome i, K is a constant of proportionality, and H_i is the degree of uncertainty (or surprise) one assigns to outcome i. The base of the logarithm is rather unimportant and may be regarded as affecting only the constant K.

It is only natural to ask why one invokes logarithms. There are two interrelated justifications for using logarithms: one mathematical, the other heuristic.

Axiomatically, there are certain utilitarian properties that a valid measure of uncertainty should possess. It is convenient, for example, that the measure be non-negative,

$$H(p_i) \geq 0. \tag{5.3}$$

In instances involving certainty, the measure should be decisive (i.e., indicate no residual uncertainty),

$$H(1) = 0. \tag{5.4}$$

Finally, the uncertainty about the co-occurrence of two unrelated outcomes should equal the sum of the uncertainties about the individual outcomes. Such additivity may be expressed as

$$H(p_i q_j) = H(p_i) + H(q_j), \tag{5.5}$$

where p_i and q_j are the probabilities of the independent outcomes i and j, respectively. It can be formally demonstrated (Aczel and Daroczy, 1975) that only the logarithmic functional form can simultaneously satisfy these three requirements.

The utility of the logarithmic form is best illustrated by a property related to its additivity, namely, the logarithmic function neatly quantifies the number of factors in the statement of the question that generates the uncertainty. To see what this means, it helps to consider a decision tree

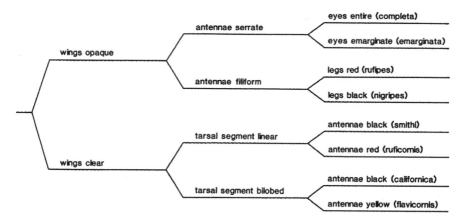

FIGURE 5.1. Hypothetical binary key to separate eight insect species (Mayr, 1969).

familiar to most biologists—a binary taxonomic key. Here taxa are "keyed" according to whether or not they possess certain attributes (e.g., vertebrate vs. invertebrate, compound eye vs. simple eye, etc.) or exceed certain demarcation parameters (e.g., greater or fewer than three spine rays, longer or shorter than 20 cm, etc.). If each key or descriptor serves as a node on a branching tree (Figure 5.1), it is easy to see that with 10 keys one can generate descriptions for 1024 distinct taxa. If nothing is known about the distribution of these species, then the probability of the first organism sampled belonging to taxon i is 1/1024. The uncertainty about this outcome according to (5.2) is 10 K (using base 2 logarithms), reflecting the 10 keys used to create the classification. Conversely, to "key out" an organism, one must answer exactly 10 questions before becoming certain about the identity of the animal.

This almost trivial example serves to demonstrate that uncertainty increases as the number of possible combinations (results) grows. Of course, the number of possible combinations increases geometrically with the number of categories used to describe the classification. However, the logarithm of a factorial or a power is a sum or product, respectively. Hence, the logarithm of the number of possible combinations reveals the number of factors responsible for generating the perceived uncertainty.

Discussion so far has focused on the uncertainty pertaining to a single outcome. A suite of outcomes is possible, however, and it is useful to quantify the uncertainty about all possible results. The probabilities of each category (the p_i's) provide convenient weighting factors for averaging the uncertainties. For, if f_i is any characteristic of category i and p_i is the probability of outcome i, then the expectation value of f, denoted by $\bar{f}$, is defined as

$$\bar{f} = \sum_i p_i f_i. \tag{5.6}$$

That the expectation value is an estimator of the average value of a sample is shown in Example 5.1. In particular, f_i may be taken as the uncertainty about outcome i (see [5.2]), and the average uncertainty (H) may then be written as

$$H = \sum_i p_i H_i, \tag{5.7a}$$

or

$$H = -K \sum_i p_i \log p_i. \tag{5.7b}$$

Example 5.1

The weights in kilograms of 70 adult males are classified into 11 intervals as listed in Table 5.1. The weights of the 70 individuals add up to 5052 kg, giving an average weight of 72.17 kg. The median of each category is shown in the column headed by f_i. The probabilities for each category (p_i) are estimated by dividing the number of individuals in each category (x_i) by the total number in the sample. The expected weight ($\bar{f}$) calculated according to (5.6) is 72.36 kg. In this particular case, the expected weight was only about 3% greater than the actual mean.

TABLE 5.1. Weights (in kg) of 70 individual males separated into 11 categories (i).

i	Range	f_i	x_i	p_i	Individual weights (kg)						
1	96–100	98	1	.0143	98						
2	91–95	93	1	.0143	91						
3	86–90	88	3	.0429	86	86	87				
4	81–85	83	11	.1571	83	82	81	82	84	85	81
					83	85	82	81			
5	76–80	78	12	.1714	77	76	80	76	76	78	79
					78	78	77	80	79		
6	71–75	73	14	.2000	73	73	71	74	75	74	72
					71	72	73	74	72	73	75
7	66–70	68	10	.1429	69	66	68	68	66	70	70
					68	67	69				
8	61–65	63	8	.1143	61	61	64	63	62	65	64
					64						
9	56–60	58	6	.0857	57	60	57	58	56	56	
10	51–55	53	3	.0429	55	53	53				
11	46–50	48	1	.0143	49						

If one knows nothing about the probabilities associated with a given set of categories, then usually each category is assumed to be equiprobable; that is, $p_i = 1/n$ for all categories $i = 1, 2, \ldots, n$. This assumption is known as Bernoulli's Principle of Insufficient Reason. The average uncertainty from (5.7b) is then

$$H_{\max} = -K \sum_i (1/n) \log (1/n), \qquad (5.8a)$$

or

$$H_{\max} = K \log n \qquad (5.8b)$$

Uncertainty is intuitively greatest under maximal ignorance; and it may be rigorously shown that $H_{\max}$ is the upper limit on H (Aczel and Daroczy, 1975).

5.3 Information

The association of a positive upper bound with a state of total ignorance about the distribution of outcomes is a significant achievement. For now any knowledge about the distribution of outcomes can serve only to decrease ignorance and, concomitantly, the average uncertainty about the results. The appropriate measure of the information is quite naturally the consequent decrease in average uncertainty. One clearly distinguishes between uncertainty, which is directly quantified, and information, which takes on the guise of a double negative. To repeat, information is the decrease in uncertainty resulting from observation of past outcomes of similar or associated events.

To be more specific, with no a priori knowledge, the starting uncertainty is maximal ($H_{\max}$). Measurements are then undertaken to count the frequencies of outcomes in various categories (x_i). After a suitably large number of observations have been made, the probabilities of the various categories can be estimated as $p_i = x_i/m$, where m is the total number of observations. The a posteriori uncertainty has been reduced to H as calculated from (5.2). The information inherent in the m observations becomes $H_{\max} - H$. H is the residual uncertainty after making the m observations. It remains to resolve this uncertainty, as best possible, through future measurements.

One should avoid regarding H as the information inherent in the distribution of p_i (as is done so often in the ecological literature). One may view the uncertainty as synonymous with diversity or complexity. One may also regard H as an upper bound on the information that may be obtained from future experiments. But to consider H as information per se can be confusing and highly misleading.

Example 5.2

If one initially knew nothing about the weights of the individuals in Example 5.1, the a priori uncertainty according to (5.8b) would be

$$H_{max} = K \log 11.$$

For the sake of this example, the base of the logarithm is chosen to be 2 and K to be unity. H_{max} then works out to be 3.4594 bits of uncertainty. One "bit" is the amount of uncertainty associated with a single binary decision.

After weighing the 70 individuals, the a posteriori uncertainty may be calculated from (5.7b) as in Table 5.2, where the p_i are taken from Table 5.1. The decrease in average uncertainty becomes $3.4594 - 3.0352 = .4242$ bits of *information*, or about .0061 bits/datum.

Actually the uniform distribution of Bernoulli ($p_i = 1/n$) is only an extreme example of an a priori distribution. It often happens that one has a vague idea of the actual distribution. For example, there are very few adult males weighing over 120 kg, and likewise only rare individuals weigh less than 50 kg. A knowledge of these limits is helpful in setting up the categories. Furthermore, some initial assumption might be made concerning the distribution within these categories (normal, exponential, etc.); and this guess (p_i^*) can be used to reduce the starting uncertainty below H_{max}. Subsequent measurements reveal a refined distribution (p_i). The assumed distribution implies an uncertainty $-K \log p_i^*$ for category i. The a posteriori uncertainty for i is $-K \log p_i$. The decrease in uncertainty (information) about outcome i is thus,

TABLE 5.2. Spreadsheet calculation of the average uncertainty (H) according to (5.7b).

i	p_i	$-\log_2 p_i$	$-p_i \log_2 p_i$
1	.0143	6.1278	.0876
2	.0143	6.1278	.0876
3	.0429	4.5429	.1949
4	.1571	2.6702	.4195
5	.1714	2.5446	.4361
6	.2000	2.3219	.4644
7	.1429	2.8069	.4011
8	.1143	3.1291	.3577
9	.0857	3.5446	.3038
10	.0429	4.5429	.1949
11	.0143	6.1278	.0876

$$-\sum_i p_i \log_2 p_i = 3.0352$$

$$(-K \log p_i^*) - (-K \log p_i), \qquad (5.9a)$$

or

$$K \log (p_i/p_i^*). \qquad (5.9b)$$

The average (expected) decrease in uncertainty is calculated by identifying (5.9b) with f_i in (5.6) to yield the average information gained as

$$I = K \sum_i p_i \log (p_i/p_i^*). \qquad (5.10)$$

Whereas the individual terms (5.9) may be positive or negative, their sum, the average information gained (5.10), will always be non-negative. This is due to the concavity of the logarithmic function (Aczel and Daroczy, 1975). At worst, further measurements will add no new information (i.e., the initial distribution will remain unchanged). Any change in the distribution always results in an overall gain in information.

Example 5.3

In addition to the 70 body weights reported in Example 5.1, another 930 individuals were subsequently classified according to the 11 weight categories. The summary of the results for the 1000 individuals is shown in the column headed x_i in Table 5.3, with the corresponding p_i's in the next column. The a priori frequencies (from Example 5.1) appear in the column labelled p_i^*. In this instance, the p_i are refinements on the p_i^*, with the initial sample of 70 being only marginally adequate for the law of large numbers to take effect. The additional information garnered from the 930

TABLE 5.3. Spreadsheet calculation of the decrease in average uncertainty obtained by weighing an additional 930 individuals.

i	Range (kg)	x_i	p_i	p_i^*	$p_i \log_2 (p_i/p_i^*)$
1	96–100	10	.010	.014	−.005
2	91–95	21	.021	.014	.012
3	86–90	62	.062	.043	.033
4	81–85	130	.130	.157	−.035
5	76–80	178	.178	.171	.010
6	71–75	195	.195	.200	−.007
7	66–70	176	.176	.143	.053
8	61–65	120	.120	.114	.009
9	56–60	71	.071	.086	−.020
10	51–55	31	.031	.043	−.015
11	46–50	6	.006	.014	−.007
					$I = .028$

weights is calculated from (5.10) and turns out to be only .028 bits. (An average of .00003 bits/datum!)

The results agree well with intuition. The initial 70 measurements are rich in information; that is, quite effective in reducing uncertainty. (An average of .0061 bits/datum.) More of the same type of data becomes progressively less efficacious in reducing uncertainty. To resolve H any further, data on associated phenomena must be sought.

Thus far, information has been created only by repeated measurements of the same phenomenon. Example 5.3 suggests that further measurements of the same type provide diminishing gains in information, and one is often left with a residual uncertainty (which may be rather large). Thereafter, one must probe associated events to see if the new phenomena can resolve the remaining uncertainty. For example, by analyzing records of a certain fish harvest over a number of years, one perceives that the salinity in the breeding grounds 2 years prior to the harvest may have an impact upon the catch in a given year. How much does a knowledge of the associated variable (the salinity) decrease the uncertainty (provide information) about the primary variable (the harvest of fish)? To answer this question, it becomes necessary to define two new concepts— joint probability and conditional probability.

The joint probability, as its name implies, is the probability that two distinct events occur together or in a given sequence. For example, if b_i represents a fish harvest falling into category i and a_j represents a salinity 2 years prior in category j, then $p(a_j, b_i)$ represents the probability that a_j occurs and is followed exactly 2 years later by b_i. One normally tabulates joint occurrences in matrix form, as in Example 5.4.

If the joint probability is normalized by the overall probability that a_j occurs, the result is the conditional probability,

$$p(b_i|a_j) = p(a_j, b_i)/p(a_j). \tag{5.11}$$

One may think of $p(b_i|a_j)$† as an adjusted probability; that is, prior to any knowledge of a_j, the estimate of b_i occurring was $p(b_i)$. After it is known that a_j has occurred, the estimate of b_i happening is revised to $p(b_i|a_j)$. Equation (5.11) is referred to in some texts as Bayes' Theorem.

Example 5.4

Three categories of clam harvest in Chesapeake Bay are designated by b_1, b_2, and b_3 such that

---

† The vertical bar used to indicate conditional probabilities should not be confused with the diagonal slash denoting division.

$$b_1 > 3000 \text{ MT}$$

$$b_2 \ 1500\text{-}3000 \text{ MT}$$

$$b_3 < 1500 \text{ MT},$$

while four classifications are used to represent the minimum January air temperatures near the midpoint of the bay 1 year prior to each harvest,

$$a_1 < -12°C$$

$$a_2 \ -10 \text{ to } -12°C$$

$$a_3 \ -8 \text{ to } -10°C$$

$$a_4 > -8°C.$$

The joint occurrences of these conditions may be regarded as the entries into what is known as a cross-classification table. Hypothetical data for 50 seasons of harvest might look like:

	a_1	a_2	a_3	a_4
b_1	9	3	3	1
b_2	2	6	5	2
b_3	2	3	4	10

One can easily convert these frequencies of joint occurrence into joint probabilities by dividing each entry by the total number of observations (50):

		a_1	a_2	a_3	a_4	$p(b_i)$
	b_1	.18	.06	.06	.02	.32
$p(a_j, b_i)$	b_2	.04	.12	.10	.04	.30
	b_3	.04	.06	.08	.20	.38
	$p(a_j)$	.26	.24	.24	.26	1.00

The matrix of joint probabilities has been extended to show the overall or marginal probabilities. The last column contains the row sums of the joint probabilities $p(b_i) = \Sigma_j p(a_j, b_i)$; whereas the last row contains the column sums, $p(a_j) = \Sigma_i p(a_j, b_i)$.

To obtain the conditional probabilities $p(b_i|a_j)$, each entry in the joint probability matrix is divided by the column sum $p(a_j)$, as in (5.11).

		a_1	a_2	a_3	a_4
	b_1	.69	.25	.25	.08
$p(b_i|a_j)$	b_2	.15	.50	.42	.15
	b_3	.15	.25	.33	.77

These tables could be used to forecast clam harvests (a very risky proposition with only 50 years of data!). Not knowing the January temperature during the preceding year, one could only say that the probability of having a good harvest this year would be $p(b_1) = .32$. However, if it were known that the previous January temperature averaged $-12.7°C$ (a_1), then the probability of a good harvest could be revised upwards to $p(b_1|a_1) = .69$. Similarly, the a priori probability of having a moderate harvest is $p(b_2) = .30$; whereas an observed January temperature of $-6°C$ (a_4) adjusts this forecast down to $p(b_2|a_4) = .15$. Under warm January conditions a low harvest is most probable, $p(b_3|a_4) = .77$.

Once the meaning of the conditional probability is clear, it becomes a straightforward matter to calculate the information about b_i provided by a knowledge of a_j. The a priori uncertainty about b_i is given by (5.2) as $-K \log p(b_i)$. After a_j is known, that uncertainty can be changed to $-K \log p(b_i|a_j)$. The decrease in uncertainty upon knowing a_j is,

$$[-K \log p(b_i)] - [-K \log p(b_i|a_j)] \tag{5.12a}$$

$$= K \log p(b_i|a_j) - K \log p(b_i) \tag{5.12b}$$

$$= K \log [p(b_i|a_j)/p(b_i)]. \tag{5.12c}$$

The information (decrease in uncertainty) represented by (5.12c) is not positive for every pair of occurrences, i and j. But when each term is weighted by the corresponding joint probability, one obtains a non-negative quantity called the average mutual information,

$$A(b; a) = K \sum_i \sum_j p(a_j, b_i) \log [p(b_i|a_j)/p(b_i)] \tag{5.13}$$

(McEliece, 1977). A represents the amount of the original uncertainty (H) about b_i, which is resolved by a knowledge of a_j. As one might expect, H serves as an upper bound on A, so that

$$H \geq A \geq 0. \tag{5.14}$$

Knowledge of another factor (a_j) can never increase the original uncertainty about b_i. At worst, it will provide no information ($A = 0$) or decrease H by an insignificant amount. (The statistical significance of A is addressed by Kullback, 1959.)

Example 5.5

To calculate how much uncertainty about the clam harvest of Example 5.4 was resolved by knowing the January water temperature, one begins by calculating the initial uncertainty

$$H = \sum_{i=1}^{3} p(b_i) \log_2 p(b_i)$$

$$= -[.32 \log_2(.32) + .30 \log_2(.30) + .38 \log_2(.38)]$$

$$= 1.58 \text{ bits}$$

Calculation of the decrease in uncertainty (A) from (5.13) is outlined in Table 5.4. Actually, only .27 bits, or 17% of the original uncertainty about the harvest, has been resolved by measuring the January temperatures.

The interpretations of a_j and b_i can be varied. In communication theory, for example, a_j represents the sending of the jth cipher, and b_i represents the reception of the ith cipher. (The reader should convince himself that the maximum amount of information is transmitted when the character received is always the same as the one sent!) Some authors try to fit all problems into the communications format; for example, salinity is a "signal" sent by the environment and the fish harvest is the "message" received by the fishery. However, such procrustean measures are unnecessary, if not misleading. Information theory is a universal calculus and pertains wherever repeated measurements are being made. An infinite variety of combinations of a_j, b_i are possible: the amino acid in the jth entry of a nucleotide sequence with the ith phenotypic characteristic; a component of momentum in the jth spatial segment of a fluid flow with the corresponding component in the ith segment; a specific political action by a given country towards nation j with the change in posture of nation i; expenditures by sector j of an economy with the income of sector i; the

TABLE 5.4. Spreadsheet calculation of the average mutual information about the clam harvest in Example 5.4 that accrues from knowing the January water temperature during the previous year.

i, j	$p(a_j, b_i)$	$p(b_i \vert a_j)$	$p(b_i)$	$p(a_j, b_i) \log_2 p(b_i \vert a_j)/p(b_i)$
1,1	.18	.69	.32	.20
1,2	.06	.25	.32	−.02
1,3	.06	.25	.32	−.02
1,4	.02	.08	.32	−.04
2,1	.04	.15	.30	−.04
2,2	.12	.50	.30	.09
2,3	.10	.42	.30	.05
2,4	.04	.15	.30	−.04
3,1	.04	.15	.38	−.05
3,2	.06	.25	.38	−.04
3,3	.08	.33	.38	−.02
3,4	.20	.77	.38	.20
				$A = .27$ bits

efflux of a given material from species j of an ecosystem with the intake by species i of the same community, etc.

The last two examples are especially germane, as they pertain to flow networks. How well does a network of exchanges function as a matrix for communication among the nodes? Specifically, how much, on the average, does an event at one compartment of the network engender action at another specific node? These issues bear heavily upon the idea of development, and it remains to show that A is a very appropriate tool to address this last question.

5.4 Summary

To clearly grasp the nature of information, it is necessary to distinguish between uncertainty and information. The uncertainty about the outcome of a given event is measured directly by the logarithm of the probability of that outcome. Further measurements that refine the probability estimates of the possible outcomes will, on the average, decrease the uncertainty. Information is characterized by the magnitude of this decrease in uncertainty. Refinements in probability estimates may derive either from further observations on the given phenomenon or be the result of data on associated events. In the latter case, the degree of coherence between the two phenomena is characterized by a quantity called the average mutual information.

6
The Description

> "If, as we have been led to think, self-determination is the criterion of growth, and if self-determination means self-articulation, we shall be analyzing the process by which growing civilizations actually grow if we investigate the way in which they progressively articulate themselves."
>
> Arnold J. Toynbee
> *A Study of History*

6.1 The Network Perspective

With the rather long prologue now over, it is time to begin the promised description of growth and development. In Chapter 5, it became clear that the success of information theory is owed to the fact that the logarithmic form of the entropy function is unique in fulfilling certain desiderata. The verbal, intuitive concepts are most fundamental and are given concrete quantitative expression in the consequent mathematical definitions. Information theory is built around a superb mathematical definition that assigns a number to common notions.

In a larger sense, the laws of thermodynamics are also definitions. While formulating the first law, thermodynamicists were creating the modern concept of energy. In the guise of the second law, the everyday sense of irreversibility took on an objective, quantitative meaning. The task at hand is to formulate a sound, quantitative definition for growth and development.

In line with the perspective advanced in Chapter 2, the definition should be macroscopic and phenomenological. It must be formulated at the level of the whole system and be valid regardless of the mechanisms at smaller scales. It will be phenomenological insofar as it is a codification or synthesis of real, observed phenomena. Any statement made under these restrictions is also likely to be universal in scope; that is, it should apply not only to ecosystems, but to every system wherein autonomous behavior might occur—ontogenetic, economic, meteorological, sociological, or otherwise. Such criteria suggest that the description will resemble a thermodynamic principle.

Although one is restricted to considering growth and development only in the most general terms, those terms, nonetheless, can be quite meaningful. The purpose of Chapter 3 was to convince the reader that flow is a sufficiently general concept pertaining to transformations throughout nature. Dynamic systems of practically every ilk can be interpreted as flow networks. Furthermore, measurements of flows indirectly pertain to the myriad of composite phenomena that leave their trace upon the fluxes. For example, organism behavior, morphology, genetic composition, and variations in physical environment all influence the members of an eco-system flow network, and their descriptions are considered to be embed-ded in the measurements of the flows.

6.2 Growth

Now that the object of narration is flow networks, the question naturally arises as to what it means for a network to grow and develop. It is argued here that growth and development are actually the extensive and inten-sive aspects of a single process. The more straightforward characteristic is growth. Growth usually implies increase or expansion, and increase may occur either in spatial extent or as accretion of the medium of flow. How do these two manifestations of increase appear in networks?

It was mentioned in Chapter 3 that the spatial configuration of an eco-system could be treated by dividing the domain into small segments (gridworks in most instances). Each segment may be considered as a node in the network, and physical transport between the segments will be represented by arcs connecting spatially adjacent nodes. Expansion of the spatial domain thus increases the number of nodes. If the ecosystem is spatially homogeneous, then each node will represent an ecological entity (e.g., species or trophic level), so that an increase in the number of nodes implies that new entities are created by some process, for example, speci-ation or migration. If one is studying n ecological components distributed over m spatial segments, the total number of nodes in the associated network will be $m \times n$. In both cases, the number of compartments in a network is one measure of its "size," so that an increase in that number reflects an aspect of growth.

Regarding the accretion of material, it is clear from Chapter 3 that the limelight here is upon the amount of medium flowing through a given compartment, and that compartmental throughput may characterize the "size" of that node. The size of the entire system becomes the sum of the individual throughputs, or the total system throughput. Growth, in this sense, translates into an increase in the total system throughput. This is really not as foreign as it first sounds. In economics, the size of an eco-nomic community is measured by its gross national product (GNP), which is a component of the total system throughput; so that in practically everyone's mind, the growth of a national economy has become synony-mous with an increase in the GNP.

In summary, the two parameters characterizing the growth of a flow network are the number of compartments and, perhaps more importantly, the total systems throughput.

6.3 Development

Characterizing development requires a little more reflection. In the dictionary, most definitions of the word overlap with those aspects of growth considered in the last section. But the intention at this point is to focus on the property of development that is independent of the size of the system. As an expedient, one may define development to be an increase in organization. While this device dispenses with size, it immediately prompts the question: "What is meant by organization?"

At this point, it pays to be literal. Webster (*New Collegiate Dictionary*, G.C. Merriam Co., 1981) defines the verb "to organize" as "to arrange or form into a coherent unity or functioning whole." This is practically synonymous with one meaning of the verb "to articulate," as "to form or fit into a systematic whole." It is revealing that articulation also means clear and precise communication, for in a very real way, flows are avenues of communication. It makes sense, then, that in a highly organized system, a signal (flow) issuing from one compartment should engender an effect at another site over a highly articulated (jointed and unambiguous) pathway. Expressed in other words, if one knows that a flow has left compartment i at time t, then in a highly organized system, this provides a great deal of information about which compartment j will receive the message (flow) at time $t + \theta$.

Examples 5.4 and 5.5 indicate how one might quantify the degree of articulation inherent in a flow network. There, the average mutual information (5.13) was used to quantify how much, on the average, knowing the air temperatures during the previous year tells one about the magnitude of the coming year's clam harvest. The same index can be used to measure the information about the inputs to the compartments of a network, which results from knowing the outputs from each compartment an instant earlier; that is, how well articulated (unambiguously linked) or organized the network of communication (flows) is.

Figure 6.1 helps to illustrate the last statement (see also Rutledge et al., 1976; Hirata and Ulanowicz, 1984). Here compartment O represents the source of all exogenous inputs at time t (i.e., the origin of the D_i), whereas components $n + 1$ and $n + 2$ are exogenous sinks to receive medium leaving the system at time $t + \theta$. Flows to $n + 1$ may still be utilized by other systems (the exports, E_i), while fluxes directed to $n + 2$ are dissipated (as the respirations, R_i).

The quantity $p(a_j)$ represents the probability that an arbitrary quantum of medium leaves compartment j at time t; and $p(b_i)$ is the corresponding probability that any quantum enters component i at time $t + \theta$ shortly thereafter. Equation (5.13) is invoked to measure the information of the flow structure; that is, how well, on the average, the network articulates a flow event from any one node to effect any other specific locus.

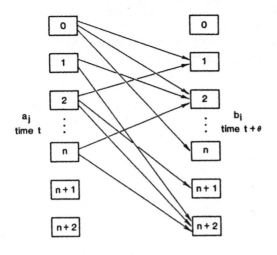

FIGURE 6.1. Temporal representation of flows among the compartments of a system. Compartment 0 represents the source of all exogenous inputs, $n + 1$ the sink for all useable exports and $n + 2$ the sink for all dissipations. Boxes on the left represent system nodes at time t and those on the right depict the same nodes at time $t + \theta$.

$$A(b; a) = K \sum_{i=0}^{n+2} \sum_{j=0}^{n+2} p(a_j, b_i) \log[p(b_i|a_j)/p(b_i)] \qquad (6.1)$$

It remains to estimate the probabilities in (6.1) in terms of the actual flows. Recalling that the measured flow from compartment j to i is T_{ji}, the sum of all the outputs from compartment j becomes $T_j = \Sigma_{i=0}^{n+2} T_{ji}$ (see 3.3). The sum of all the inputs to compartment i will be denoted by $T_i' = \Sigma_{j=0}^{n+2} T_{ji}$ (as in 3.2). The total systems throughput will be written simply as $T = \Sigma_{j=0}^{n+2} T_j = \Sigma_{i=0}^{n+2} T_i'$. (Note that $T_{oi} = D_i$, $T_{j,n+1} = E_j$, and $T_{j,n+2} = R_j$.) The probability $p(a_j)$ will be denoted by Q_j and is estimated by

$$Q_j = p(a_j) \sim T_j/T. \qquad (6.2)$$

Similarly,

$$Q_i' = p(b_i) \sim T_i'/T. \qquad (6.3)$$

The only way for a quantum to both leave j and enter i is for it to be part of the flow T_{ji}, therefore the joint probabilities should be estimated as

$$p(a_j, b_i) \sim T_{ji}/T. \qquad (6.4)$$

Now by (5.11), the conditional probability is the quotient of (6.4) by (6.2),

$$p(b_i|a_j) = p(a_j, b_i)/p(a_j) \sim T_{ji}/T_j, \qquad (6.5)$$

but the quotient in (6.5) is recognized as the host coefficient from Chapter 3,

$$T_{ji}/T_j = f_{ji}. \qquad (6.6)$$

This conveniently allows one to write the joint probability as the product of the host coefficient and Q_j,

$$p(a_j, b_i) = p(b_i|a_j)\, p(a_j) \sim f_{ji}Q_j. \tag{6.7}$$

Finally, substituting (6.3), (6.5), (6.6), and (6.7) into (6.1) gives the average information inherent in the network in terms of measurable quantities, namely:

$$A(b; a) = K \sum_{i=0}^{n+2} \sum_{j=0}^{n+2} f_{ji}Q_j \log(f_{ji}/Q'_i) \tag{6.8}$$

Example 6.1*

Calculate the value of the average mutual information for the Cone Spring network (Figure 3.2) in units of K.

The host coefficients have already been calculated in Example 3.5, but the entire procedure is repeated here for clarity. The five compartments and the three boxes representing the rest of the universe generate an 8×8 matrix of flows.

	0	1	2	3	4	5	6	7	RS
0	0	11,184	635	0	0	0	0	0	11,819
1	0	0	8881	0	0	0	300	2003	11,184
2	0	0	0	5205	2309	0	860	3109	11,483
3	0	0	1600	0	75	0	255	3275	5205
4	0	0	200	0	0	370	0	1814	2384
5	0	0	167	0	0	0	0	203	370
6	0	0	0	0	0	0	0	0	0
7	0	0	0	0	0	0	0	0	0
CS	0	11,184	11,483	5205	2384	370	1415	10,404	42,445

The row sums appear in the last column of the table and the column sums are in the last row. The total system throughput appears in the lower right corner. Because the system is balanced (in steady-state), the column sums equal the row sums for compartments 1–5. Also, the column totals for compartments 6 and 7 add up to the row sum for compartment 1, indicating that the whole system is balanced. The host coefficients (f_{ji}) are found by dividing each entry by its corresponding row sum.

---

* Appendix B contains an aid to calculating the principal information indices introduced in this chapter.

	0	1	2	3	4	5	6	7
0	0	.946	.054	0	0	0	0	0
1	0	0	.794	0	0	0	.027	.179
2	0	0	0	.453	.201	0	.075	.271
3	0	0	.307	0	.014	0	.049	.629
4	0	0	.084	0	0	.155	0	.761
5	0	0	.451	0	0	0	0	.549
6	0	0	0	0	0	0	0	0
7	0	0	0	0	0	0	0	0

The Q_j and Q'_i are the row sums and column sums, respectively, after normalization by T.

$$
(Q) = \begin{pmatrix} .278 \\ .263 \\ .271 \\ .123 \\ .056 \\ .009 \\ 0 \\ 0 \end{pmatrix}, \quad (Q') = \begin{pmatrix} 0 \\ .263 \\ .271 \\ .123 \\ .056 \\ .009 \\ .033 \\ .245 \end{pmatrix}
$$

All the elements in (6.8) are now at hand. When substituting into (6.8), any $f_{ji} = 0$ makes no contribution to the mutual information. This is a consequence of the fact that the limit of $x \log x$ as x goes to zero is itself zero. The terms in (6.8) may be displayed in matrix form and summed over all rows and columns to obtain the average mutual information. Logarithms to the base 2 are used in these calculations.

	0	1	2	3	4	5	6	7	
0	0	.486	−.035	0	0	0	0	0	.451
1	0	0	.325	0	0	0	−.002	−.021	.302
2	0	0	0	.231	.100	0	.024	.011	.366
3	0	0	.007	0	−.003	0	.003	.105	.112
4	0	0	−.008	0	0	.036	0	.070	.098
5	0	0	.003	0	0	0	0	.006	.009
6	0	0	0	0	0	0	0	0	0
7	0	0	0	0	0	0	0	0	0
	0	.486	.292	.231	.097	.036	.025	.169	1.336

Hence, $A = 1.336$ bits of K.

Example 6.2

The three closed flow networks in Figure 6.2 all have identical total system throughputs (96 units). The network in Figure 6.2a is maximally connected. Each compartment exchanges medium with all compartments in equal amounts. Although one might know that a particle is leaving a certain compartment, this knowledge yields no information about which node is likely to receive it as input. On the average, one is highly uncertain about the consequences of any flow. The configuration is poorly articulated, as reflected by the value $A = 0$ that one obtains by substituting the appropriate flow values into (6.8).

The network in Figure 6.2b is somewhat better articulated. Each component contributes flow equally to only two of the remaining nodes. Once it is known that a quantum is exiting from a given locus (e.g., 3), two of the remaining nodes (2 and 3 in this case) may be excluded as the destination. When substitution is made into (6.8), A is found to equal 1 unit of K. (Because logarithms to base 2 were used in this calculation, the value 1 reflects the single binary decision, which on the average has been resolved by knowing which pair of flows is allowed and which is prohibited.)

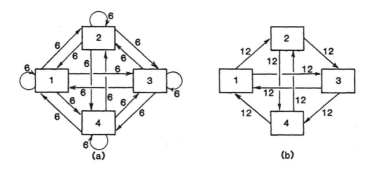

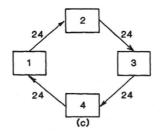

FIGURE 6.2. Three artificial, closed networks with differing degrees of articulation. (a) The maximally connected and minimally articulated configuration of 96 units of flow connecting four nodes. (b) An intermediate level of articulation. (c) The maximally articulated configuration of flows.

Finally, the network in Figure 6.2c is maximally articulated. Knowing the box from which a particle departs unequivocally reveals its destination. The formula in (6.8) yields $A = 2$ units of K, which is the largest value possible for a network with four nodes. (Again, the value 2 reflects the two binary decisions needed to specify the destination of the one existing flow leaving each node.)

6.4 Simultaneous Growth and Development

As mentioned earlier, growth and development are both aspects of a unitary process, although the English language appears to lack a single word encompassing both factors. Thus far, the total systems throughput has been seen to characterize the size of a flow system, but it conveys no sense of how the system is organized. Conversely, the average mutual information quantifies the organization of a flow network, but does not sufficiently address system size.

Now the average mutual information is scaled by the constant K, which has been assiduously retained in all references to information variables. The usual convention is that K defines the units of information. For example, if the base of the logarithm is 2, a single unit of K is referred to as 1 "bit" (as in Example 5.2). Should natural logarithms be used, $K = 1$ then represents one "nat" of information; when the logarithmic base is 10, K is measured in "hartleys." Early in most introductions to information theory, the base of the logarithms is specified; K is set equal to 1, and thereafter it disappears from discussion. However, Tribus and McIrvine (1971) suggest that the purpose of K is to impart physical dimensions to the index it scales. As the total systems throughput has already been cited as characterizing the size (or scale) of a network, it is appropriate to equate K with T. Accordingly, (6.8) becomes,

$$A = T \sum_{i=0}^{n+2} \sum_{j=0}^{n+2} f_{ji} Q_i \log(f_{ji}/Q_i). \tag{6.9}$$

All of the elements of size and organization appear in (6.9). Size (the extensive element) is represented by T and the number of compartments n, while the system structure (the intensive factor) is determined by the Q_i and f_{ji}. Organization takes the form of mutual information. Henceforth, to refer to the amalgamation "size and organization" as symbolized in (6.9), the term "ascendency" will be used. "*Growth and development*," having been identified as an increase in network size and organization, therefore, translates into an *increase in ascendency*.

The word "ascendency" was chosen for the dual interpretation it affords. The overt connotation of domination or supremacy reflects the advantages of size and organization. To win out over other systems (real or putative), an entity must have a propitious combination of size and organization. Size alone will not guarantee success, as was the case with Goliath. Conversely, it is difficult to imagine, even in this day of powerful and sophisticated weaponry, that a small but highly developed state such as Luxembourg could by itself long fend off the hostile advances of a superpower. The intention here is to use (6.9) to quantify the ascendency exercised by the combination of size and organization. However, the root of the word ascendency lies with the verb "to ascend" or "to rise." By definition, the ascendency of a growing and developing network will increase. The structure appears to arise from the mire of chaos, irrespective of any reference to competition or control.

6.5 Ascendency Arising From a Dynamic Tension

It is a very old idea that the order in the world results from a struggle between countervailing forces; e.g., good vs. evil, Ying vs. Yang, etc. Without implying any specific analogies, growth and development (or increasing ascendency) can also be seen to arise from a tension between two seemingly opposing tendencies. To see this, it is helpful to rewrite (6.9) in a different form (suggested by H. Hirata). The quantity Q_j may be interjected into the numerator and denominator of each argument in the expression for system ascendency.

$$A = T \sum_{i=0}^{n+2} \sum_{j=0}^{n+2} f_{ji} Q_j \log(f_{ji} Q_j / Q_j Q_i'), \qquad (6.10)$$

and the logarithmic contribution may be separated into three terms,

$$A = T \sum_{i=0}^{n+2} \sum_{j=0}^{n+2} f_{ji} Q_j \log(f_{ji} Q_j) - T \sum_{j=0}^{n+2} \left[\sum_{i=0}^{n+2} f_{ji} \right] Q_j \log Q_j$$

$$- T \sum_{i=0}^{n+2} \left[\sum_{j=0}^{n+2} f_{ji} Q_j \right] \log Q_i'. \qquad (6.11)$$

Looking at the terms in brackets,

$$\sum_{i=0}^{n+2} f_{ji} = \sum_{i=0}^{n+2} T_{ji}/T_j = T_j/T_j = 1, \qquad (6.12)$$

and

$$\sum_{j=0}^{n+2} f_{ji} Q_j = \sum_{j=0}^{n+2} (T_{ji}/T_j)(T_j/T) = \left(\sum_{j=0}^{n+2} T_{ji} \right) \Big/ T = T_i'/T = Q_i', \quad (6.13)$$

so that (6.11) may be rewritten as

$$A = -T \sum_{j=0}^{n+2} Q_j \log Q_j - T \sum_{i=0}^{n+2} Q_i' \log Q_i'$$

$$- \left[-T \sum_{i=0}^{n+2} \sum_{j=0}^{n+2} f_{ji} Q_j \log(f_{ji} Q_j) \right]. \tag{6.14}$$

All three terms in (6.14) have the form of an entropy (5.7b). The first two terms are called the output entropy and the input entropy, respectively, reflecting the fact that Q_j is defined in terms of outputs (see 6.2) and Q_i' in terms of inputs (see 6.3). They are both non-negative quantities. The term in brackets is named the joint entropy, because it is comprised of joint probabilities (see 6.7). The entire third term is always nonpositive and is generated by the interconnections between the compartments (the f_{ji}).

The first two terms in (6.14) represent the uncertainty about the distribution of flows among the nodes. Various processes such as speciation, immigration, and random disturbances serve to increase both of these terms as time passes. Similar influences augment the last term; that is, the spontaneous appearance of arbitrary links or stochastic changes in existing links would, in the absence of any selection pressure, ultimately increase the joint entropy. In fact, if one were to assemble compartments and connect them in a random way, the expectation value of the joint entropy would exactly balance the input and output entropies, as can be seen in Figure 6.2a. However, any nonrandom selection (e.g., positive feedback as discussed in Chapter 4) would retard, or possibly even reverse, any increase in the joint entropy and thereby would increase the system ascendency.

In most previous treatments of evolution, attention has been focused upon only those processes that serve to diminish the joint entropy by streamlining the topology or increasing the efficiencies of the nodes. However, one must also recognize the potential that disordering events afford for growth and development by virtue of their contributions to increasing the input and output entropies. To rephrase Atlan (1974), disordering events continually provide the new variety that can be woven into a new structure to meet the exigencies of a changing environment. Conrad (1983) describes such adaptability at length and strongly emphasizes its necessity for the evolution of systems.

6.6 Generic Limits to Growth and Development

Another ancient truism is that nothing may increase without bound. Everything that grows is constrained by temporal, spatial, or material factors. Obviously, such constraints also serve to keep system ascendency within limits. To see how restrictions on A arise, it is helpful to split (6.10) into only two terms,

$$A = -T \sum_{j=0}^{n} Q_j \log Q_j - \left[-T \sum_{i=0}^{n+2} \sum_{j=0}^{n+2} f_{ji} Q_j \log (f_{ji} Q_j / Q_i') \right], \quad (6.15)$$

where the term in brackets is a non-negative quantity called the conditional entropy, because it measures the uncertainty remaining after the flow structure has been specified. By writing the non-negative quantity A as the difference between two other inherently non-negative quantities, it becomes apparent that the first term in (6.15) serves as an upper bound on A. This quantity will be assigned the symbol C and called the development capacity,

$$C = -T \sum_{j=0}^{n} Q_j \log Q_j, \quad (6.16)$$

and it is evident from (5.14) that

$$C \geq A \geq 0. \quad (6.17)$$

It follows that limits on the growth of C will also be limits on the increase of A. In turn, the two factors that limit C are T (the total systems throughput) and n (the number of compartments) (e.g., 5.8b).

The total systems throughput (T) is ultimately limited by the total amount of inputs. Although T may increase via greater cycling or through transfers to a larger number of compartments, each transfer must have its associated cost. The effect of these obligatory losses is that flow beginning a cycle will always be attenuated by the time it completes the circuit (e.g, Example 4.1). The residual flow after m passes through the cycle will be proportional to the mth power of the attenuation factor. Because the attenuation factor is less than 1, summing the flows over an infinity of serial passes through the loop results in an infinite series (a scalar version of [3.11]), which converges to a finite limit. Hence, recycling per se cannot increase T without bound.

A greater number of compartments would serve to increase the sum in (6.16); and, in fact, a proliferation of species during the early stages of development is often observed. However, there are practical limits to this trend. As the available flows of medium become distributed over more components, the average throughput per compartment must decrease. Inevitably, some of the compartments will possess throughputs that are so small as to make them highly vulnerable to extinction by random perturbations. The rigors of the environment, therefore, eventually preclude division of the community into an unlimited number of elements.

Conversely, a more benign, less stochastic environment allows finer partitioning, and thus a greater development capacity per given amount of input. The most familiar ecological example of a high development capacity is the tropical rain forest, where the physical environment is relatively more predictable. Although sources of medium are spare, a large degree of recycling inflates T and a low level of perturbation permits a high degree of partitioning.

Example 6.3

Calculate the development capacity for the Cone Spring ecosystem network.

The Q_i and total system throughput for Cone Spring were calculated in Example 6.1.

$$T = 42,445 \text{ kcal m}^{-2} \text{ y}^{-1}, \quad (Q) = \begin{pmatrix} .278 \\ .263 \\ .271 \\ .123 \\ .056 \\ .009 \\ 0 \\ 0 \end{pmatrix}.$$

Substituting into (6.16) one gets

$$C = T \times \left(- \sum_{i=0}^{n} Q_i \log Q_i \right)$$

$$= 42,445 \times (2.1966)$$

$$= 93,232 \text{ kcal bits m}^{-2} \text{ y}^{-1}$$

In Section 2.2, thermodynamic work was described as an ordering process. The ascendency A takes on the same units as those of work when the medium in question is energy. The development capacity C then appears as an upper bound on work in the same sense that the energy does in (2.5). Just as not all the energy of a system can be made available for work, so not all the development potential can appear as organized flow structure. A certain fraction of C, represented by the second term in (6.15), must always be encumbered for other reasons. The second term in (6.15), the conditional entropy, will be referred to hereafter as the systems overhead, Φ. It is obvious that the magnitude of systems overhead is another constraint on the increase in A.

Because there are four categories of flow in a real network (inputs, intercompartmental transfers, exports and dissipations), the overhead term Φ may be divided into four separate contributions. From Figure 6.1, it is clear that $f_{jo} = f_{n+1,i} = f_{n+2,i} = 0$, out of which

$$\Phi = -T \sum_{i=0}^{n+2} \sum_{j=0}^{n+2} f_{ji} Q_j \log(f_{ji} Q_j / Q_i)$$

$$= -T \sum_{i=1}^{n} f_{oi} Q_o \log(f_{oi} Q_o / Q_i')$$

$$-T \sum_{i=1}^{n} \sum_{j=1}^{n} f_{ji} Q_j \log(f_{ji} Q_j / Q_i')$$

$$-T \sum_{j=1}^{n} e_j Q_j \log(e_j Q_j / Q_{n+1}')$$

$$-T \sum_{j=1}^{n} r_j Q_j \log(r_j Q_j / Q_{n+2}'), \qquad (6.18)$$

where $e_j \ (= E_j / T_j)$ and $r_j \ (= R_j / T_j)$ have been used in place of $f_{j,n+1}$ and $f_{j,n+2}$ to characterize the exports and respirations, respectively. Because of the relationship $\sum_{j=0}^{n} f_{ji} Q_j = Q_i'$, it may be immediately deduced that the argument of each logarithm in (6.18) is less than or equal to 1, implying that the components of Φ are all termwise non-negative. The four terms in (6.18) are denoted by Φ_o, Φ_r, Φ_e, and Φ_s, respectively.

Example 6.4

Calculate the various terms in the overhead of the Cone Spring network of Example 6.1.

All of the quantities necessary to evaluate (6.18) have already been computed in Example 6.1. The results of the substitution are

$$\Phi_o = 2652 \text{ kcal bits m}^{-2} \text{ y}^{-1} \ (2.8\% \text{ of } C)$$

$$\Phi_r = 10{,}510 \text{ kcal bits m}^{-2} \text{ y}^{-1} \ (11.3\% \text{ of } C)$$

$$\Phi_e = 1920 \text{ kcal bits m}^{-2} \text{ y}^{-1} \ (2.1\% \text{ of } C)$$

$$\Phi_s = 21{,}364 \text{ kcal bits m}^{-2} \text{ y}^{-1} \ (22.9\% \text{ of } C).$$

Total overhead is 36,446 kcal bits m^{-2} y^{-1} or 39.1% of C.

One way to increase A is to decrease Φ. Looking at the four components of Φ in turn, it is obvious that the coefficient of T in Φ_o will be minimized when all the inputs are concentrated into one compartment. This tendency is tempered by the fact that total inputs serve as a bound on T, so that it becomes possible that additional inputs to other compartments might increase T more than the consequent spread of inputs increases Φ_o. The interior functional redundancy, Φ_r, is minimized when only one input enters each compartment, or only one flow exits each compartment, and when the input flows balance the outputs. Then the argument of each logarithm in Φ_r becomes unity, so that $\Phi_r = 0$. The export and respiration terms behave similarly, and both are minimized by a tendency to focus all activity into one compartment. Such a concentration of export activity is, in principle, possible; but the second law demands that nonzero losses of throughput occur in each compartment. Hence, Φ_s is necessarily greater than zero.

Example 6.5

The hypothetical network in Figure 6.3 resembles a classical administrative hierarchy. The full development capacity is 158.3 flow bits. Of this amount, 62.4% appears as ascendency and 37.6% as dissipative overhead Φ_s. The remaining components of the overhead, Φ_r, Φ_o, and Φ_e are all identically zero. In particular, the absence of interior functional redundancy (Φ_r) makes this configuration well-suited for communication.

To many, this topology epitomizes the idea of organization. It is interesting to note, however, that the dissipative term encumbered by tree-like structures very often is a large fraction of the development capacity. Therefore, it should come as no surprise that organizations with recycle or feedback often outcompete rigid hierarchies.

Thus, the magnitude of Φ is seen to be sensitive to network topology. To better understand ascendency, it is helpful to consider the topologies of networks with minimal overhead (maximal ascendency). For this purpose, it is temporarily useful to assume that dissipation can be made vanishingly small. Without loss of generality, one may suppose that $T =$

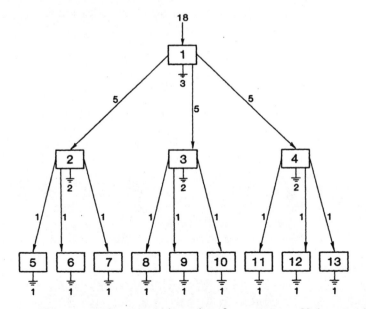

FIGURE 6.3. Flows among a nested hierarchy of components. Units are arbitrary.

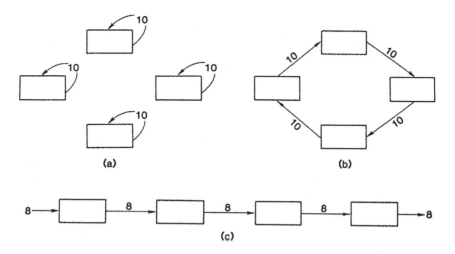

FIGURE 6.4. Three canonical configurations of 40 units of flow in a four-component system. Each topology is maximally articulated: (a) Four separate self-feedback loops. (b) A single feedback loop incorporating all components. (c) A chain of transfers.

40 units of flow and $n = 4$ compartments. The development capacity will be maximized by equal distribution of the throughput among the compartments. To minimize the overhead, there should be single inputs and single outputs from each component. While there are numerous configurations that would realize the full development capacity, all such networks could be considered combinations of the three elementary topologies illustrated in Figure 6.4.

In the topology in Figure 6.4a, the self-loops might represent storage (see Section 3.4), and the configuration would be static (or at equilibrium in the thermodynamic sense). Such a structure (or absence of structure) is neither open nor living in the common sense of the word. It is worth mentioning, however, that there may be conditions under which an ever-increasing, overall ascendency will become dominated by the exogenous transfers and will lead to the breakup of the system or to the equilibrium point (death).

The configuration in Figure 6.4c looks like a trick! There is a phantom fifth compartment not shown in the diagram representing the external world. The straight chain has a higher ascendency than the other two structures (464 units as compared with 320 for both Figures 6.4a and 6.4b) by virtue of its being an open system. Acyclical, nonautonomous networks (trees in the language of graph theory) can be thought of as adjunctions of such straight pathways. When a physical system is constrained away from equilibrium, the tendency for increasing ascendency could drive the system towards a configuration of least dissipation (Prigogine, 1947), which usually resembles a tree.

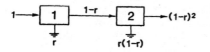

FIGURE 6.5. A primitive chain of two processes, each of which transfers medium with an efficiency $1 - r$.

Example 6.6

In the extremely simple network in Figure 6.5, the two elements are assumed to process medium at the same efficiency, $1 - r$, where r is the respiration coefficient. Without loss of generality a total input of 1 unit enters the first compartment, and the amount $(1 - r)^2$ is exported from the second compartment. How does the ascendency change as the rate of respiration r decreases; that is, as the compartments become more efficient?

From the diagram one reads the throughputs

$$T_0 = 1 \qquad T_0' = 0$$
$$T_1 = 1 \qquad T_1' = 1$$
$$T_2 = 1 - r \qquad T_2' = 1 - r$$
$$T_3 = 0 \qquad T_3' = (1 - r)^2$$
$$T_4 = 0 \qquad T_4' = r(2 - r).$$

So that

$$T = \sum_{i=0}^{4} T_i = \sum_{i=0}^{4} T_i' = 3 - r,$$

and

$$Q_0 = 1/(3 - r) \qquad Q_0' = 0$$
$$Q_1 = 1/(3 - r) \qquad Q_1' = 1/(3 - r)$$
$$Q_2 = (1 - r)/(3 - r) \qquad Q_2' = (1 - r)/(3 - r)$$
$$Q_3 = 0 \qquad Q_3' = (1 - r)^2/(3 - r)$$
$$Q_4 = 0 \qquad Q_4' = r(2 - r)/(3 - r).$$

Similarly,

$$f_{01} = 1$$
$$f_{12} = f_{13} = 1 - r$$
$$f_{14} = f_{24} = r$$

and all other $f_{ij} = 0$.

FIGURE 6.6. A graph showing the ascendency (in flow bits) of the network in Figure 6.5 as a function of the common component efficiency, $1 - r$.

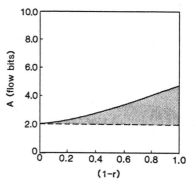

The values for T, Q_j, Q_i', and f_{ij} may now be substituted into (6.9). After some algebraic manipulation, the result is:

$$A = (3 - r) \log(3 - r) - (1 - r)^2 \log(1 - r) - r(2 - r) \log(2 - r).$$

When there is complete dissipation ($r = 1$), A becomes equal to 2 log 2. In the limit of perfect transfer ($r \rightarrow 0$), A approaches the value 3 log 3.

A glance at the derivative of A with respect to r,

$$dA/dr = 2(1 - r) \log[(1 - r)/(2 - r)] - \log(3 - r),$$

shows that this derivative is uniformly negative over the range of values of r from 1 to 0. That is, A increases uniformly as r decreases (Figure 6.6).

If one considers the first compartment as a system unto itself, A for this subsystem is a constant, 2 log 2. The difference between the solid and dotted lines in Figure 6.6, therefore, represents the increase in the system ascendency that accrues from adding another compartment. Not surprisingly, the ascendency of straight chain systems increases as both component efficiency and chain length increase. That nonautonomous cascade systems can increase in ascendency by decreasing dissipation is in accord with the Prigogine hypothesis.

It should be clear from this discussion on overhead that the relative amount of system ascendency (A/C) may be interpreted as a measure of the network "efficiency" in a dual sense of the word. Systems with a high ascendency component are usually composed of elements that are themselves thermodynamically efficient (contribute little to Φ_s) and are linked together in a streamlined fashion (as reflected by the small Φ_r).

6.7 Autonomous Growth and Development

It was implied in the discussions about Figures 6.4a and 6.4c that conditions exist under which the system ascendency can be increased only by

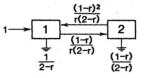

FIGURE 6.7. A rudimentary feedback loop between two compartments. As in Figure 6.5, each node has efficiency $1 - r$.

driving the overhead to a minimum. The resulting configurations are not characteristic of living systems. In contrast is the perfect feedback structure (Figure 6.4b), which in Chapter 4 was seen as a precondition for the autonomous behavior characteristic of living systems. Such feedback appeared to give rise to a selection pressure capable of ordering living networks. In addition, the growth in throughput engendered by the feedback can make internal circulation a major component of the ascendency.

Example 6.7

The export from component 2 in Figure 6.5 is shunted back into element 1, creating an internal feedback loop (Figure 6.7). The flows have been adjusted to keep the system at steady-state, while maintaining the meaning of the parameter r as the complement of the efficiency of each element. Again, one seeks to relate the behavior of A to the respiration coefficient r.

The calculation of A proceeds as in Example 6.6, and the preliminary steps are left to the reader. Towards the end of the calculation, one arrives at a form equivalent to

$$A = \log[(2 - r)(1 + r)] + \{[(1 - r)/r(2 - r)] \log[(2 - r)(1 + r)]$$
$$+ [(1 - r)^2/r(2 - r)] \log[(1 - r)(2 - r)(1 + r)]\} + \log(1 + r).$$

As before, when $r = 1$, $A = 2 \log 2$; however, A now increases without bound as r approaches zero, as shown in Figure 6.8.

Looking at the expression for A in terms of r, it is easy to trace the effects of recycle. When r is set $= 1$, the terms in braces disappear and the

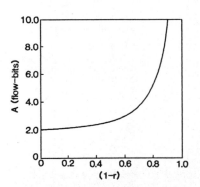

FIGURE 6.8. A graph showing the ascendency (in flow bits) of the network in Figure 6.7 as a function of component efficiency $1 - r$.

first and last terms sum to 2 log 2. The first term is generated by the input flow and the last term by the combined respiratory flows. At the other extreme, as r approaches zero, the terms in braces dominate the ascendency. These two terms are generated by the internal transfers T_{12} and T_{21} and characterize the feedback. Hence, under propitious circumstances, the internal feedback terms can come to dominate the ascendency.

If increasing ascendency depicts the growth and development of living systems, it is apparent that internal positive feedback should be implicit in the ascendency function. In order to focus upon those terms in the ascendency associated with this internal feedback, it is useful to separate (6.9) into four groups of terms (see Section 3.2) as follows:

$$A = T \sum_{i=1}^{n} f_{oi} Q_o \log(f_{oi}/Q_i')$$

$$+ T \sum_{i=1}^{n} \sum_{j=1}^{n} f_{ji} Q_j \log(f_{ji}/Q_i')$$

$$+ T \sum_{j=1}^{n} e_j Q_j \log(e_j/Q_{n+1}')$$

$$+ T \sum_{j=1}^{n} r_j Q_j \log(r_j/Q_{n+2}') \tag{6.19}$$

The first set of terms stems from flows originating outside the system and will be denoted by A_o, the second grouping (internal exchanges) by A_I, the third (useful exports) by A_e, and the last (dissipations) by A_s, so that

$$A = A_o + A_I + A_e + A_s \tag{6.20}$$

It is possible to further decompose the key component A_I in a manner similar to that following (6.11).

$$A_I = T \sum_{i=1}^{n} \sum_{j=1}^{n} f_{ji} Q_j \log(f_{ji} Q_j / Q_j Q_i') \tag{6.21}$$

$$A_I = T \sum_{i=1}^{n} \sum_{j=1}^{n} f_{ji} Q_j \log(f_{ji} Q_j / Q_i') - T \sum_{j=1}^{n} \left[\sum_{i=1}^{n} f_{ji} \right] Q_j \log Q_j \tag{6.22}$$

Now

$$\sum_{i=1}^{n+2} f_{ji} = 1 \text{ for all } j, \text{ so that}$$

$$\sum_{i=1}^{n} f_{ji} = 1 - e_j - r_j. \tag{6.23}$$

Replacing the term in brackets by (6.23) and rearranging terms gives

$$A_I = -T \sum_{j=1}^{n} Q_j \log Q_j$$

$$-\left\{ -T \sum_{j=1}^{n} e_j Q_j \log Q_j \right.$$

$$-T \sum_{j=1}^{n} r_j Q_j \log Q_j$$

$$\left. -T \sum_{i=1}^{n} \sum_{j=1}^{n} f_{ji} Q_j \log (f_{ji} Q_j / Q_i') \right\} \tag{6.24}$$

The first term in (6.24) looks exactly like the development capacity, except that it is missing the term for the external inputs ($j = 0$). It may appropriately be called the internal development capacity C_I, and it shares an upper bound with the total capacity, C. The first term in braces is non-negative and is caused by the export of useable material to other systems. It is given the name "tribute" and the symbol E. The second term in brackets is identical in form to the tribute, but stems from the dissipative losses. Dissipation is represented by the letter S (in lieu of the conventional sigma used in irreversible thermodynamics to denote dissipation). The final term is assuredly non-negative only at steady-state ($Q_i = Q_i'$ for all $i = 1, n$). It is a measure of the ambiguity of the internal connections within the system. As this ambiguity mostly results from the multiplicity of pathways between two arbitrary compartments, the term is characterized as the functional redundancy of the network (Rutledge et al., 1976) and is given the symbol R. (Note from [6.18] that $R = \Phi_r$.) The functional redundancy will be referred to hereafter simply as the "redundancy," but it should not be confused with the same term often employed in communications theory, where redundancy can mean something quite different.

6.8 The Limits to Autonomous Growth and Development

The internal ascendency (A_I) markedly resembles the full ascendency. For example, (6.24) when rewritten as

$$A_I = C_I - (E + S + R), \tag{6.25}$$

has the same form as (2.5) and (6.15) and first appeared in Ulanowicz (1980). The sum of the terms in parentheses is called the internal overhead.

However, the limits that the internal overhead places upon the internal ascendency are more rigorous than those that Φ imposes on the increase in A. For example, looking at the forms of S and E, it is clear that *any* losses to the external world detract from the internal ascendency. A little algebra will show that

$$E = \Phi_e + A_e \tag{6.26}$$

and

$$S = \Phi_s + A_s. \tag{6.27}$$

Since Φ_e and Φ_s are both non-negative, it follows that

$$E \geq A_e \tag{6.28}$$

and

$$S \geq A_s. \tag{6.29}$$

In other words, losses always subtract from the internal capacity of the system more than they ever contribute to the overall ascendency.

Example 6.8

Calculate the internal capacity, the internal ascendency, and the components of the internal overhead for the Cone Spring ecosystem of Example 6.1.

As with Example 6.4 the calculation is simply a matter of substituting the appropriate factors into the defining equations.

The internal capacity is the first term in (6.24) and is the full capacity minus the contribution from the inputs . $C_I = 71,372$ kcal bits m^{-2} y^{-1}. The internal ascendency from (6.21) is 29,332 kcal bits m^{-2} y^{-1}, or 41% of the internal capacity. The redundancy was calculated in Example 6.4 as 10,510 kcal bits m^{-2} y^{-1}, and it now constitutes 14.7% of the internal capacity. The dissipation, the third term in (6.24), is 40% of internal capacity, or 28,558 kcal bits m^{-2} y^{-1}. Finally, the tribute is a mere 4.2% of internal capacity, or 2971 kcal bits m^{-2} y^{-1} as calculated from the second term in (6.24). The fact that the internal ascendency is barely more than half the overall ascendency indicates that the system is strongly driven by its inputs and losses, as one might expect of a flowing spring ecosystem with heavy advective exchanges.

Just as it was impossible to vanquish Φ for any real system, it is even more difficult to decrease the elements of the internal overhead towards zero without, at some point, adversely impacting the living system. For example, the second law mandates that each component contribute to the dissipation S. While at first this may seem like a pure waste, it should be remembered that much of this dissipation occurs for the purpose of building structure (i.e., doing work) at lower hierarchical levels.

The limits to reducing the dissipation are not clear. The smallest dissipation rates occur near thermodynamic equilibrium. However, near-equilibrium throughput is so small that system size (as measured by total system throughput) diminishes to near extinction. In order to create significant ecological flow structure, appreciable dissipation must occur.

This tension between efficiency and rate of productivity was noted by Lotka (1922), who suggested that the organisms' rate of productivity might be more important to their survival than the efficiency at which production is carried out. Lotka's idea has been championed in recent times by H.T. Odum (Odum and Pinkerton, 1955; Odum, 1971), who extends the notion to the ecosystem level. Ecosystems are said to evolve so as to maximize the power produced; that is, they maximize the total systems throughput, T. Maximum power production also has been applied to development in microbiological (Westerhoff et al., 1983) and engineering systems (Andresen et al., 1977).

It would be exciting to report that the trade-off between production and dissipation is implicit in the optimization of the overall ascendency. That is, perhaps the A_l and A_r components achieve a compromise that optimizes the overall ascendency. Unfortunately, there are no explicit reasons why this generally might be the case. Defining the explicit constraints on dissipation in general terms remains one of the most important and intriguing unfinished tasks in irreversible thermodynamics. Odum argues that r_i should rise with increasing compartmental throughput, and such a trend makes intuitive sense for asymptotically large throughputs. However, it is also reasonable to conceive of very small compartments (i.e., those with small T_i), the efficiency of which would benefit from increased throughput. It is sufficient to remark here that there exist as yet unspecified thermodynamic restraints on how small compartmental dissipation may become without jeopardizing the continued existence of that compartment.

If dissipation represents the system-level effect of transfers to lower hierarchical levels, then its complement is in the tribute E, which characterizes the given system's contribution to larger composite systems. (This hierarchical distinction between S and E is precisely why pains were taken earlier to differentiate losses as comprising either export or dissipation.) To what extent may system tribute be minimized? The answer depends upon the nature of the larger system. If within the larger network, there is no communication from the output of the given subsystem back to its own inputs (i.e., if the subsystem participates in no cycles within the larger system), then eliminating the exports is in the best interest of increasing subsystem ascendency. However, should the subsystem engage in feedback within the larger ensemble, then increasing the subsystem ascendency by reducing its tribute will become a negative perturbation to the larger cycle. It will not be rewarded; and, as discussed in Chapter 4, the larger cycle(s) will select against such perturbation to the detriment of the offending subsystem.

Examples of curbs on reducing tribute are not too difficult to imagine. Suppose that on a certain island, the fruits and seeds of local flora are food for migratory birds. It might appear to augment the island ecosystem ascendency if those exports were internalized, for example, through the appearance of local species of birds or insects that prey on the fruits. However, if the migratory birds bring locally scarce trace elements or nitrogen from distant sources, diverting fruit production to local predators would be counterproductive to the flora of the island, and hence to the entire island ecosystem.

As another hypothetical example, within the global economic network, a given state is assumed to be extremely rich in a certain necessary resource (e.g., oil), but otherwise does not possess other internal resources in abundance. If this state attempts to cut back on exporting the scarce resource to drive up the price of this commodity beyond reasonable limits, the economy of the rest of the world might suffer to an extent that inevitably would impair the economy of the exporting state. Such was the aftermath of the 1970s oil crisis.

While the restraints on diminishing S and E lie at other scales of the hierarchy, the same cannot be said for the limits to reducing the redundancy R, which is a topological property existing entirely at the level of the given system. A nonzero value of R indicates the presence of multiple pathways between at least some pairs of system components. In the absence of constraints, growth along the "most efficient" internal pathways would streamline the network and increase ascendency at the expense of redundancy. What keeps systems from achieving their most efficient, streamlined states?

The reason lies in the fact that implicit in R is the capacity of the system to respond to disturbance. Supposing that it were possible for a given system to ascend to a state of zero redundancy, then the resulting streamlined network would be highly vulnerable to unexpected perturbations. A straight chain or a perfect cycle is a very rigid structure, and a disturbance at a single point can have catastrophic consequences for the overall organization. However, as Odum (1953) pointed out, when multiple connections between components exist, it then becomes possible for a disturbance along one path to be compensated to some degree by a countervailing change in flow along a less affected parallel pathway. Odum's suggestion eventually led to the hypothesis mentioned earlier that "diversity begets stability," but it would be counterproductive to be drawn into a discussion on these terms. In fact, May (1973) argued rather convincingly that the proposition is not *generally* true.

The difficulty with the diversity-homeostasis argument is that it is an insufficient portrayal of reality. In one sense, it appears diversity does induce homeostasis. At other times, however, it seems a homeostatic environment permits a high diversity of flows (May, 1973). Diversity and homeostasis should be regarded as a cybernetic couple, each influencing the other. The interplay between these two attributes is perhaps best described in the context of the larger struggle between the ordering and disordering tendencies evident in (6.14). On one hand are the entropic, disordering events that serve to increase the R of any real network. The opposing tendency is for autonomous systems to grow and develop (increase in ascendency). Any real structure results from the balance of these countervailing pressures. Near that balance, the relative amounts of A and R possessed by a system are a reflection of rigors of the environment. If the balance is upset by an extraordinary perturbation, a temporary excess of R in relation to A will ensue. The subsequent recovery (growth of A at the expense of R) may be termed resiliency. Should a system overshoot the balance point through development during a period of unusual calm, it then becomes all the more vulnerable to the next perturbation that reverses matters. Hence, a rigorous environment should encumber a large fraction of redundancy; a more benign environment, a lesser fraction. All real environments encumber a nonzero value of R.

It appears as though a simpler, complete description of homeostasis is not possible. For example, there is truth in the statement that a large redundancy buffers against disturbance by allowing compensation. Thus, there is scant possibility that the extinction of one arbitrary species of insect in a tropical rain forest will set the entire ecosystem to crumbling. However, the "redundancy begets stability" argument neglects why a large value of R exists in the first place. The benign environment of the rain forest permits the development capacity to attain a high level. There is sufficient capacity so that even if A takes up a large fraction of C, the absolute value of R may still be rather large. If the situation is to continue, however, any perturbation had better not be so destructive as to degrade the "benigness" of the overall environment. Again, reality is seen to involve circular causality. Any description of part of the cycle is by nature incomplete.

The association of R with the capacity for resilience raises a provocative question. If the recovery of a system to perturbation is contingent upon network configurations that contribute to R, then are not these structural attributes also aspects of the system's organization? That is, it appears as though the ascendency does not characterize all facets of network organization as was intended.

It is important to remember that ascendency is a phenomenological index. It is impossible to quantify a system at all scales or in every conceivable dimension. Insofar as one chooses to define and quantify a network, ascendency will characterize all those aspects of organization that can be made explicit by the actual measurements. Those processes commonly associated with organization, but which are beyond measurement, may contribute to the overhead and can possibly serve to limit the observed ascendency. In the latter instance, organization at one level is being constrained by organization at another hierarchical level or dimension ("hierarchical compensation" in Conrad, 1983).

To make matters a little clearer, it has already been mentioned that some dissipation may be necessary to build and maintain structures at sub-compartmental levels. By way of contrast, it was argued how tribute might be necessary to sustain larger structures, of which the quantified system is but a component. In both cases, organizational processes may lie at scales where measurement is either impossible or has been precluded. Widening the scope of quantitative resolution to allow measurement of these hitherto obscure processes will make explicit their contribution to the ascendency calculated from the enlarged set of data.

The issue of redundancy and homeostasis goes beyond hierarchical considerations and involves the added dimension of time. To fully quantify the response of a system to perturbation, it becomes necessary to measure temporal changes in the network structure and the external world. Thus far, only steady-state networks have been discussed, but in Section 7.3, the ascendency of dynamic systems will be addressed. Suffice it to say here that once the dimension of time has been included in the quantitative description of a network, the temporal adaptations of that system to changing conditions will appear explicitly in the revised ascendency.

Of course, not all overhead consists of "hidden" organization. Expanding the resolution of a network will "explain" some, but never all, of the overhead. The second law of thermodynamics is no chimera. (See also, the notion of "indifference" in Conrad, 1983.)

To summarize the previous two sections, the effect of internal feedback on ascendency is concentrated in a component called the internal ascendency. When autonomous behavior occurs, the internal ascendency will form a significant and possibly dominant fraction of the increase in overall ascendency. The previous description of growth and development, when confined to autonomous systems, may be modified to read: *An autonomous system behaves over an adequate interval of time to optimize the internal network ascendency subject to thermodynamic, hierarchical, and environmental constraints.* These constraints are similar to those on general systems, except that system losses appear more prominently.

The process of growth and development has now been cast in the familiar form of a variational principle (see Section 2.4). When the optimization problem is actually implemented as an algorithm (see Section 7.7), the objective function (the ascendency) is *maximized* subject to a set of conservation constraints. Strictly speaking, such maximization is to be carried out only over the immediate neighborhood of the instantaneous or past system configuration. The system engages, so to speak, in "local hill climbing," proceeding as far up the local incline of ascendency as conditions will allow (Allen, personal communication). The community need not be presumed to discern any distant global maximum towards which it is striving. In recognition of the local nature of the maximization process, and for lack of a better term, the variational statement in the previous paragraph will be referred to as the "principle of *optimal* ascendency."

A characteristic of any variational principle is that it often connotes an image of competition. Now competition is surely part, although not all, of the picture of growth and development. But does it make any sense to talk of an ecosystem competing as a unit? After all, only one ecosystem can occupy a given spatial domain, and that community realizes a single evolutionary pathway. How does one interpret a variational statement under such circumstances?

Actually, variational principles have long been applied to physical systems under similar circumstances. For example, a solid body thrown into the air traverses a single determinate pathway. It is possible, however, to imagine other trajectories that differ from the real pathway by an infinitesimally small amount. For each of these pathways, real or imagined, one may compute an attribute called "the Hamiltonian." It can be shown (Goldstein, 1950) that the Hamiltonian of the observed trajectory is always greater than that for any nearby putative pathway. The flying body behaves as if it maximized its Hamiltonian with respect to any nearby imaginary trajectory. Hence, it is not necessary to speak about ecosystems engaging in competition (although such a situation should not be precluded). Rather, it is sufficient merely to state that the actual course of system development engenders greater ascendency than any imaginable nearby pathway of network evolution.

Hamilton's portrayal of solid body mechanics affords no real advantage over the more popular treatment of the subject, which employs Newton's second law in differential equation form. Hamilton's integral statement and Newton's differential law are demonstrably equivalent. However, it is argued here that such parity of description does not exist across the analogous hierarchical boundaries of biology. Most ecological modelers take a "microscopic" approach to their subject in assuming rigid topologies for the network of system interactions and fixed functional descriptions of component processes. Such methods are regarded here as insufficient to provide an adequate picture of any autonomous behavior that may exist at the level of whole ecosystems. The variational statement of optimal ascendency, by contrast, was consciously formulated as a macroscopic proposition, so that it will be capable of describing emergent behavior.

6.9 Phenomenological Basis for Optimal Ascendency

Thus far, ascendency has been introduced in epistemological fashion, as if one could derive the principle of optimal ascendency by reason alone. The advantage of such a presentation is that it quickly leads one to discern exactly what is being said. It should come as no surprise, however, that this was not the manner in which ascendency was first formulated; for if it were, the title of this book would indeed be ill-considered. Phenomenological descriptions should begin instead as highly intuitive efforts to weave incomplete, disconnected, and sometimes contradictory observations into a single theoretical fabric.

In all candor, the phenomenological basis for this treatise is largely second-hand. Observations on the attributes of developing systems made over the last 50 years have been conveniently summarized by Odum (1969) in his seminal article, "The strategy of ecosystem development." Odum listed 24 attributes of ecosystems that he believed were correlated with development. The listing of these properties is reproduced as Table 6.1.

Unfortunately, not all 24 criteria are readily interpreted as pertaining to flow networks. It is necessary, therefore, to take some license in translating several of Odum's remarks into flow terminology. For example, stocks may generally be read as compartmental throughputs (as in Section 3.4).

Three of the attributes In Table 6.1 (11, 12, and 24) come very close to saying that A and A/C tend to increase during development. The obverse of this trend is that the fraction of development capacity encumbered by overhead (Φ/C in 23 and A_e/C in 4) progressively decreases in the course of maturation.

Example 6.9

A hypothetical simple network is shown in Figure 6.9. In relation to the simpler network in Figure 6.10, the compartments in 6.9 are "generalists," whereas those in 6.10 are "specialists." In rerouting the flow in 6.10, an effort was made to keep the respiration quotients of the four compartments nearly the same as in 6.9. The rerouting resulted in a slight (5.3%) increase in the total systems throughput from 525 to 553, and an even smaller increase in development capacity (3.7%). The ascendency, however, rose by some 22% from 594.8 to 726.7 flow bits; while the overhead fell by 23% from 417.9 to 323.3 flow bits, mostly because of a decline in redundancy.

Three of the attributes in Table 6.1 (8, 9, and 10) focus on development capacity, C. It has already been discussed how increasing the number of compartments and the partitioning of flows augments the upper bound on ascendency.

Example 6.10

The network in Figure 6.11 is a modification of 6.9, where compartment 4 has been split into two parts. While the total system throughput has thereby increased only 1.3% to 531.7 flow units, the corresponding increase in development capacity was 7.3%. Ascendency rose by an even greater amount (10.7%). (This greater increase in ascendency is due mainly to the specialist nature of the new component.)

TABLE 6.1 A tabular model of ecological succession: trends to be expected in the development of ecosystems.

Ecosystem attributes	Developmental stages	Mature stages
Community energetics		
1. Gross production/community respiration (*P/R* ratio)	Greater or less than 1	Approaches 1
2. Gross production/standing crop biomass (*P/B* ratio)	High	Low
3. Biomass supported/unit energy flow (*B/E* ratio)	Low	High
4. Net community production (yield)	High	Low
5. Food chains	Linear, predominantly grazing	Web-like, predominantly detritus
Community structure		
6. Total organic matter	Small	Large
7. Inorganic nutrients	Extrabiotic	Intrabiotic
8. Species diversity-variety component	Low	High
9. Species diversity-equitability component	Low	High
10. Biochemical diversity	Low	High
11. Stratification and spatial heterogeneity (pattern diversity)	Poorly organized	Well organized
Life history		
12. Niche specialization	Broad	Narrow
13. Size of organism	Small	Large
14. Life cycles	Short, simple	Long, complex
Nutrient cycling		
15. Mineral cycles	Open	Closed
16. Nutrient exchange rate, between organisms and environment	Rapid	Slow
17. Role of detritus in nutrient regeneration	Unimportant	Important
Selection pressure		
18. Growth form	For rapid growth ("*r* selection")	For feedback control ("*K* selection")
19. Production	Quantity	Quality
Overall homeostasis		
20. Internal symbiosis	Undeveloped	Developed
21. Nutrient conservation	Poor	Good
22. Stability (resistance to external perturbations)	Poor	Good
23. Entropy	High	Low
24. Information	Low	High

From Odum (1969).

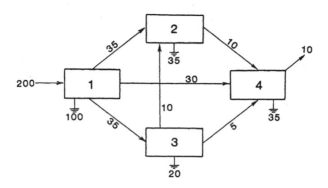

FIGURE 6.9. A fabricated network of flows among four components.

By far the largest number of Odum's attributes (2, 3, 7, 15, 16, 17, 20, and 21) can be construed in some way to imply that mature systems exhibit more cycling and greater internalization of medium. That is, the system tends to conserve medium both by storing it in the components and by cycling it within the system. Now, adding flows to the internal transfers serves to boost the total systems throughput. It also tends to augment the internal fraction of the development capacity (C_I/C) by making the compartmental Q_i's ($i = 1, 2, . . . , n$) larger at the expense of inputs and losses. Providing the augmented cycles do not substantially increase the redundancy (R), the internal ascendency should also rise. Thus, increased cycling does have the potential for enhancing T, C, and A_I.

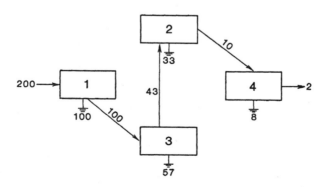

FIGURE 6.10. A network of flows simplified from the one in Figure 6.9 by rerouting all flow along the pathway 1-3-2-4. Each component behaves more as a specialist in comparison to its counterpart in Figure 6.9.

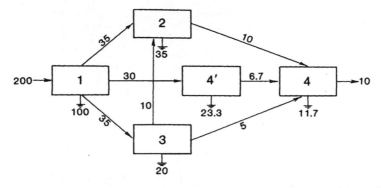

FIGURE 6.11. A modification of the network in Figure 6.9 created by splitting component 4 into two separate nodes.

Example 6.11

In the network in Figure 6.12, the former export from compartment 4 (see Figure 6.9) is shunted back into compartment 3 in such a way as to hold all the compartmental r_i nearly constant. Two internal cycles are thereby created; the first from 3 to 4 and return, and the second from 3 to 2 to 4 and back to 3.

The augmented cycling increases the total system throughput from 525 to 540 flow units, the development capacity from 1013 to 1070 flow bits, the internal capacity even more from 734 to 782 flow bits, and the internal ascendency from 144 to 154 flow bits.

One would expect, then, that internalization and cycling would usually contribute to the *overall* ascendency of a system. With all other things being equal, this is not generally the case! The overall ascendency for the

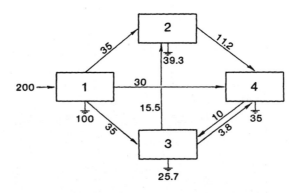

FIGURE 6.12. The network in Figure 6.9 is modified here by shunting the export from node 4 back into node 3. The former export flow is thereby "internalized" and two new cycles are created.

network in Figure 6.12, for example, was 14.8 flow bits *less* than the *A* of Figure 6.9. It appears that *arbitrary* cycling and feedback frequently produce more systems overhead than ascendency. Pimm (1982) and May (1983) remark upon the rarity of cycles in the *feeding* webs of most ecosystems. How, then, does one reconcile Pimm's observations with Odum's emphasis on cycling, all within the context of optimizing overall ascendency? Clues may be found in Example 6.7.

In order to be effective, cycling needs to be a reasonable fraction of the total system activity. Also, it helps if the cycling reduces the effective losses from the system. Now, if cycling is to involve an appreciable fraction of the total throughput, it must pass through the larger, lower trophic compartments. Furthermore, if the effective losses are to be lowered, it would help if the key compartments exhibited low dissipation. Both of these criteria are met by the nonliving, "ground-level" components associated with recycle; for example, detritus. In fact, Odum cites the importance of detritus in recycle as attribute 17. What is implied, but unwritten, in Table 6.1 is that the aforementioned eight attributes dealing with cycling primarily involve detritus or other nonliving, ground-state elements. Hence, Odum's emphasis on nonliving recycle and Pimm's caution against higher trophic cycles are both consistent with the optimization of overall ascendency.

Example 6.12

Figure 4.10 is a schematic of carbon flow (mg C m^{-2} d^{-1}) among the taxa of an estuarine marsh inlet ecosystem: Crystal River, Florida.

An enumeration of the cycles in this network (Example 4.6) reveals there are 119 distinct simple cycles. All but one of those cycles passes through the detrital compartment. One wishes to assess the contribution of cycling to the ascendency of the network. Two *conservative* estimates of this effect can be made.

In the first instance, all the cycles are subtracted from the network as described in Chapter 4. The overwhelming probability is that the residual flows (Figure 4.13) are higher than what would occur in absence of the cycles (positive feedback having an amplifying effect). Nevertheless, regarding Figure 4.13 as the starting network, the addition of the cycles increases the total system throughput by 7.6% to 22,420 mg C m^{-2} d^{-1}, the development capacity by 12% to 47,086 mg C bits m^{-2} d^{-1}, and the overall ascendency by 4.3% to 28,337 mg C bits m^{-2} d^{-1}. As might be expected, the internal ascendency rises more sharply (11.7%) than the overall ascendency.

The role of detritus in the residual flows can be estimated by excising the detrital compartment from the network in Figure 4.13. The inputs to detritus from microphytes and macrophytes thereby become exports from the system. All of the subsequent direct and indirect flows issuing from

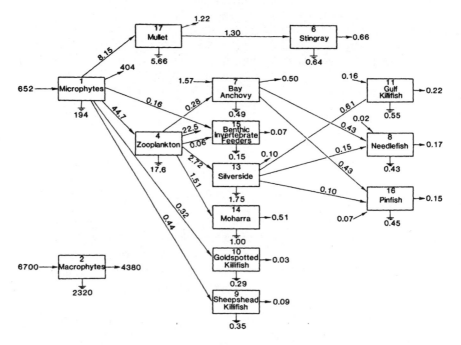

FIGURE 6.13. Residual flows (mg C m^{-2} y^{-1}) from the acyclical network in Figure 4.13 after all direct and indirect contributions from the detritus have been subtracted.

the detritus may now be subtracted from the network (thanks to the absence of cycles), and the remaining network is an estimate of what the system might look like if all the detritus were physically removed, as often happens in open water systems. The remainder appears in Figure 6.13. The removal of detritus extinguishes the benthic invertebrates and the longnosed killifish. It also separates the macrophytes from the rest of the system. Looking at only the microphyte grazing chain, the total system throughput is a mere 1369 mg C m^{-2} d^{-1}, the development capacity is down to 1779 mg C bits m^{-2} d^{-1}, and the overall ascendency comprises only 1423 mg C bits m^{-2} d^{-1}. The dominance of the macrophytes and detritus in the system is quite evident.

System size and growth form (attributes 6 and 18 in Table 6.1) are implicit in the increase of ascendency. The activity level (T) appears in (6.9) multiplying a logarithm. Now logarithmic functions are *not* particularly sensitive to changes in their arguments. As a result, during the early stages of network development, practically any mechanism that can increase the value of T will also serve to increase the overall ascendency. Hence, initial emphasis is on the rate of system growth (r selection). It is only after sources of new inputs have become scarce that further increases in ascendency can be achieved by pruning the network to create a more coherent system structure (K selection). It seems difficult to imagine how Odum's first criterion, the progression towards a balance in system production and respiration, might be implicit in an increasing network ascendency. In fact, not every change that causes a jump in A is in the direction of equilibrating production and respiration. There is good reason, however, to believe that the long term trend of increasing A leads the system toward a steady-state.

To see this, it is helpful to consider the transformation,

$$Q_i^* = Q_i', \quad i = 1, 2, \ldots n$$

$$Q_o^* = Q_{n+1}' + Q_{n+2}', \tag{6.30}$$

after which (6.9) becomes

$$A^* = T \sum_{i=0}^{n} \sum_{j=0}^{n} f_{ji} Q_j \log (f_{ji}/Q_i^*). \tag{6.31}$$

Putting Q_i into the numerator and denominator of each argument and separating terms gives

$$A^* = T \sum_{i=0}^{n} \sum_{j=0}^{n} f_{ji} Q_j \log (f_{ji}/Q_i) + T \sum_{i=0}^{n} \sum_{j=0}^{n} f_{ji} Q_j \log (Q_i/Q_i^*). \tag{6.32}$$

Recognizing that $Q_i^* = \sum_{j=0}^{n} f_{ji} Q_j$ yields

$$A^* = T \sum_{i=0}^{n} \sum_{j=0}^{n} f_{ji} Q_j \log (f_{ji}/Q_i) - \left[T \sum_{i=0}^{n} Q_i^* \log (Q_i^*/Q_i) \right]. \tag{6.33}$$

The first term in (6.33) is seen to be the ascendency the system would have if all the compartmental throughputs were determined solely by the outputs. It is inherently a non-negative quantity. The term in brackets has the same form as the relative entropy (5.10). It is non-negative and takes on the value zero if and only if $Q_i^* = Q_i$ for all i, that is, only when the system is at steady-state.

As long as the system ascendency can increase via the first term of (6.33), the network may develop despite an imbalance of medium about each compartment. However, the same constraints inhibiting increases in (6.9) should also apply to the first term in (6.33), holding that quantity below some upper bound. Further augmentation of ascendency may follow only by decreasing the second term; that is, the system is eventually driven towards steady-state. In particular, at steady-state, the production and losses are balanced (attribute 1).

Example 6.13

The network in Figure 6.9 is thrown slightly out of balance by swapping the 1–3 and 1–4 flows (as indicated in Figure 6.14), while leaving all other flows unchanged. This creates a surplus of inputs to compartment 4 and a

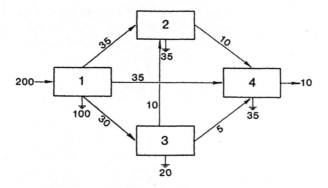

FIGURE 6.14. A permutation of the flow network in Figure 6.9 obtained by swapping the magnitudes of the 1-3 and 1-4 flows. Components 3 and 4 (and consequently the network as a whole) are placed slightly out of balance.

corresponding deficit into 3, so that the total systems throughput remains unchanged at 525 flow units; and the capacity (C) holds at 1012.7 flow bits. The overall ascendency drops, however, from 594.8 flow bits in the steady-state system to 592 in this marginally out-of-balance network.

The arguments in Section 6.7 concerning stability and homeostasis apply to Odum's 22nd attribute as well. However, (6.33) casts an interesting epilogue on the subject. For the sake of discussion, call the two terms in (6.33) U and V, respectively, so that the relation becomes

$$A = U - V. \tag{6.34}$$

Increasing A eventually drives V towards zero. If constraints are such that V may reach zero, then the steady-state configuration may persist indefinitely. If, however, the approach to $V = 0$ is somehow impeded, then part of the network will continue either to grow or to shrink until the system configuration radically changes. (Most likely, this will occur as a sudden switch or a mathematical catastrophe.)

To address the issue of persistence of system configuration, it becomes necessary to study the nature of the constraints on the system. Supposing that one arbitrarily designates the constant inputs and outputs, so that $\Sigma_{i=0}^{n} D_i \neq \Sigma_{i=0}^{n} (E_i + R_i)$. This implies that $Q_o \neq Q_o^*$. V can never equal zero, and the configuration is destined to become progressively more imbalanced. It is obviously possible to overconstrain the system.

More realistic boundary conditions might require that the external inputs, (D_i) either are given as constants or are specified as functions of time. Similarly, the losses are assumed to be known functions of time. (Under transform 6.30, it makes sense to talk only about the sum $E_i + R_i$, the total losses from each compartment.) These specifications comprise a total of $2n$ external constraints on the system. This number is to be compared with a total of $\lambda + 2n$ primitive variables comprising A, where λ is the number of internal flows (T_{ij}) and $2n$ refers to the potential number of exchanges with the universe. One now imagines that U is near its upper bound (another constraint on the problem). Is it possible to find a configuration of the T_{ij} so that $V = 0$? With $\lambda + 2n$ variables and $2n + 1$ constraints, there are $\lambda - 1$ degrees of internal freedom for the system to achieve steady-state. The number of internal flows (λ) may range from a minimum of $n - 1$ (anything less is a disconnected network) to a maximum of n^2 (everything communicates directly with everything else). Thus, the more numerous the internal flows, the more likely it becomes that the system can achieve a steady-state. However, the greater the connectivity of the compartments (λ/n^2), the higher is likely to be the redundancy (R). One is led to a recapitulation of Odum's argument for homeostasis—the greater the R, the higher the probability for a persistent configuration of the flow network. Nonetheless, one should remember (Section 6.7) that Odum's suggestion remains an insufficient description of the relationship between flow redundancy and homeostasis in ecosystems.

Three of Odum's attributes (Table 6.1) remain outside the present scope of discussion (13, 14, and 19). Should the notion of ascendency be extended to hierarchical situations, there remains the possibility that organism size, individual life cycles, and the quality of exported materials will all be amenable to description in terms of an ascendency-like function.

Criterion 5 remains problematic. Although the progression from grazing chains to detritus has already been discussed, the change from linear chains to web-like networks is not entirely consistent with the increase of ascendency during the latter stages of maturation. Whatever web-like nature an increasing ascendency might impart to a network should appear during the intermediate stages of development. During the later stages, it is more likely that transfers are being eliminated to retain only the more efficient cycles. However, the capacity of the mature system may have become so much greater than that of its pioneer configuration that even after the final decrease in connectance, the older system would appear more web-like than its early predecessor, (see also Section 7.5).

Several notable investigators have challenged the validity of many of Odum's criteria (MacMahon, 1979; Bormann and Likens, 1979). It should be obvious from what has been written in this section that there also are nontrivial discrepancies between the principle of optimal ascendency and several of Odum's measures. Inductive ideas, if they are to be any use, should point out lacunae and contradictions in existing observations—even among those that have generated the synthesis! These differences are mentioned lest anyone doubting Odum's ideas on mature communities rushes to judge the adequacy of increasing ascendency as a description of growth and development. Much remains to be told about how one might expand the notion of optimal ascendency. Besides, at least some of the disagreement between Odum and his critics is semantic in nature, and the concrete mathematical form of ascendency affords a promising vehicle with which to resolve lexical issues.

6.10 The Principle of Maximal Work

One of the primary benefits of casting a mathematical description at the level of the whole system is that it permits the quantitative definition of certain ensemble phenomena that heretofore were only vaguely imagined. A prime example is the notion of group selection. Most will contend that no evidence exists for selection at any level greater than that of two coevolving populations. Many of the same individuals would further maintain that the very idea of community-level selection is nonsensical.

These two negative assertions are not unrelated. Were it possible to define and quantify community-level forces of selection, it would be likely that empirical confirmation of such pressures would soon follow. In order to find something, it helps to have a good image of what one is searching for; otherwise, it becomes all too easy to conclude that the object of the search does not exist.

A way of quantifying selection pressure becomes apparent if one identifies the ascendency as a work function. To do this, one must choose the medium in circulation as energy (dimensions $ML^{-2}T^{-2}$). Then the dimensions of the ascendency function become power bits ($ML^{-2}T^{-3}$ bits). Augmenting the ascendency is thus related to increasing the power generated by the ecosystem, T. In fact, during the early stages of development, when T dominates the increase in A, the optimization of A is virtually indistinguishable from the Lotka maximum power principle (Odum, 1971). During the later stages of maturation, as network configuration becomes more important, the nature of ascendency as a work function becomes more evident. The resemblance of (6.15) and (6.25) to the Helmholz work function (2.5) is not accidental. In Section 2.2, the notion of work was generalized to include any ordering process. The ascendency, therefore, might be said to represent the work inherent in the creation and maintenance of order in the network flow structure. (The principle of optimizing ascendency is seen to be quite distinct from the well-known principle of minimal work. Examples of physical phenomena obeying the latter law include the trajectory of a mass within a force field or the flow of a river within a meander [Leopold and Langbein, 1966]. By contrast, the tendency towards maximal work characterizes autonomous development in systems.)

One of the properties of a work or power function is that it can be written as the sum of products of forces and flows (Section 2.4). Such a decomposition is also possible with the ascendency (Kay, personal communication). Since T is independent of either i or j in (6.9), it may be moved inside the double summations. Substituting for Q_j and f_{ji} from (6.2) and (6.6), respectively transforms (6.9) into

$$A = \sum_{i=0}^{n+2} \sum_{j=0}^{n+2} T_{ji}[\log(f_{ji}/Q_i')]. \tag{6.35}$$

In (6.35), there is one term for each system flow (T_{ji}) occurring in the network. Each flow is multiplied by a conjugate factor in brackets. Could it be that this factor corresponds to the elusive ecological force? (See discussion of forces in Section 2.4 and Ulanowicz, 1972.) Calling

$$F_{ji} = \log(f_{ji}/Q_i'), \tag{6.36}$$

the ascendency may then be written as the sum of fluxes and "pseudo-forces,"

$$A = \sum_{i=0}^{n+2} \sum_{j=0}^{n+2} T_{ji}F_{ji}. \tag{6.37}$$

There is no guarantee that the F_{ji} are positive. A positive F_{ji} would indicate a tendency to increase the associated flow, whereas a negative pseudo-force would reveal a pressure from the community to select against the conjugate flow.

The decomposition of the work function (A) does not have to be in terms of individual flows. For example, one may extend the decomposition (6.20) to identify the contribution to A from each individual compartment; that is,

$$A = \sum_{j=0}^{n+2} A_j, \tag{6.38}$$

where

$$A_j = \sum_{i=0}^{n+2} T_{ji} \log(f_{ji}/Q_i'). \tag{6.39}$$

Or alternatively,

$$A = \sum_{i=0}^{n+2} A_i', \tag{6.40}$$

where

$$A_i' = \sum_{j=0}^{n+2} T_{ji} \log(f_{ji}/Q_i'). \tag{6.41}$$

It may be shown that all A_j and all A_i' are inherently non-negative; that is, the contribution from any individual compartment to the overall work function is always positive. Does this imply that there is never any selection pressure against an arbitrarily introduced species? Certainly not! The reader need only recall the earlier assertion (6.29) that increasing the respiration of the system decreased the internal ascendency more than it added to A_r. Similarly, any species introduced into the community will make a positive contribution to the a posteriori community ascendency. The crucial comparison, however, is between the prior and subsequent community ascendencies. It is possible for decreases in contributions from the original components to more than offset any increment from the new species. If the net change in overall ascendency is positive, the new species should begin to grow as a member of the community; if it is negative, the throughput of the new species should decrease.

From the preceding argument, it appears that selection against flows might be stronger than selection against species (i.e., nodes). As an individual flow shrinks in relation to other flows out of j (i.e., f_{ji} falls), the corresponding selective pseudo-force (6.36) becomes even more negative. No similar negative feedback appears to operate against introduced species. In both situations, however, the absolute value of the term associated with either a small flow or a small compartment appears to diminish as the magnitude of the given flow or species diminishes. There comes a point at which the small changes wrought by the very small elements of the network are swamped by the larger changes in ascendency due to the major elements of the community. In the end, it is unlikely that a single species is driven to extinction by the selective forces associated with increasing ascendency. Elimination is more likely to be due to a chance environmental occurrence that extinguishes a compartment of a system. Thus, it appears possible for a large number of inconsequential species to persist in the presence of relatively few major taxa. One expects a skewed distribution of species levels not unlike the log-normal distribution observed in many ecosystems (Preston, 1948).

6.11 Relationship to Other Variational Principles

No description is complete without mentioning how it relates to others presently in vogue. It should be noted at the outset, however, that the relationships are not necessarily all competitive, as usually is the case between different reductionist hypotheses. Quite often, two phenomenological narratives will both contain substantial elements of truth, so that it is inappropriate to attempt to falsify completely one or the other principle. In fact, multiple partial descriptions are likely to be the basis for a stronger, more inclusive principle.

The principle of optimal ascendency is seen in this light to be built upon the incisive descriptions of earlier investigators. Lotka (1922) was the first to portray competition in relatively unconstrained circumstances as a contest between the magnitudes of processes, rather than between their efficiencies. Prigogine (1947), and later Katchalsky and Curran (1965), addressed transformations occurring under stronger constraints, where the very restrictions (generalized boundary conditions) become coagents with process kinetics and random perturbations in determining the system structure. The Brussels school initially emphasized structure formation (as quantified by the third term in 6.14) somewhat to the exclusion of other events occurring during development.

Ostensibly, the Lotka and Prigogine principles are antagonistic. H.T. Odum was once asked whether he thought development in ecosystems might be described by the principle of minimum entropy production. He responded by saying that any entity using minimum entropy as an adaptive strategy has a death wish! Of course, the Prigogine description has evolved well beyond merely optimizing the entropy production; but the later theory continues to address evolution under conservative circumstances, whereas adherents of the Lotka-Odum school think in expansive terms.

In mainstream ecology, this antagonism takes on the guise of r selection vs. K selection. The two aspects of development *both* occur in nature. Under conditions surrounding most immature ecosystems, an r strategy (fast growth) appears most appropriate for survival. Under the very different conditions associated with more developed systems, a K strategy (structural refinement) seems to give survival advantage. It is not a case of one strategy being correct and the other incorrect, but of separate descriptions of distinct aspects of a single growth and development process as it occurs under contrasting conditions.

Here it has been argued that growth and development should be treated as a unitary process as quantified by a single descriptor—the ascendency, which is relevant under all conditions. Calling the informational factor in ascendency W, then $A = TW$. Using the product rule for the differential of A,

$$dA = WdT + TdW. \qquad (6.42)$$

In an inchoate ecosystem, where resources have not yet been saturated, the first term on the right hand side of (6.42) dominates; and ascendency increases practically as if T were the sole index being optimized. Lotka prevails. In mature systems, where available resources are saturated, the first term diminishes, so that further increases in A occur primarily in the second term in (6.42). One way of restructuring to optimize W is to minimize the overhead, Φ. The Prigogine description is then appropriate.

Conrad (1972) has used information theory to describe the mutual development of a biological system and its physical environment. Rather than focusing upon the mutual information between the system and its universe, Conrad emphasizes the joint entropy between the two (a quantity similar to the third term in [6.14]) and says that this joint entropy is minimized during the course of development. By all appearances, he has translated the Prigogine principle into the language of information theory. Conrad's conclusions would be fortified by more concrete examples, and his notation is rather intricate, although it is tailored to address development in hierarchical systems. These points notwithstanding, there is no fundamental contradiction between Conrad's propositions and the principle of optimal ascendency. In fact, an amalgamation of the two statements should lead to an even more general description of growth and development.

Jørgensen and Mejer (1979) apply the thermodynamic concept of exergy (popular in Europe as a measure of the distance of a system from equilibrium) to ecological systems. Ecosystems are said to progress towards states of *maximal* exergy. Like the Lotka formulation, Jørgensen's description is expansive. Again, there appears to be no fundamental contradiction between maximal exergy and optimal ascendency. However, like the entropy of an ensemble (6.16), the exergy contains no explicit reference to the relationships among the components, and it is not clear whether or not the underlying structure is implicit in the exergy. Jørgensen and Mejer (1981) later point out, however, that maximal exergy was really intended to be used in conjunction with an explicit model of system structure for the purpose of estimating parameters. In this capacity, their variational principle might perform well. However, the problem remains that to calculate the exergy, it is necessary to know the entropies of the living compartments. Unfortunately, measuring the thermodynamic entropy of a living being is fraught with conceptual difficulties (see Section 2.3).

There remain several other extremal principles that are all based upon the conservative nature of ecosystems. Cheslak and Lamarra (1981) see the residence time of energy in an ecosystem as the key indicator of maturity. Residence time may increase during storage or via cycling (see also Finn, 1976). Hannon (1979) analytically reproduces Margalef's (1968) suggestion that a developing system minimizes the metabolized energy per unit of stored biomass. All such conservative behavior appears to be implicit in the optimization of internal ascendency, A_I.

At least one attempt has been made to "test" the various "hypotheses" of ecosystem development. Fontaine (1981) employed each of the available principles in an optimization program to estimate the parameters of a simulation model of Silver Spring. How closely the optimal parameters generated by each objective function approximated the measured values was purported to reflect the validity of that respective principle. Fontaine's original results pointed to maximization of power as the best objective function, although ascendency had as yet to be implemented. Later testing of the ascendency as the objective function revealed that it would reproduce the measured parameters exactly (Fontaine, personal communication). In other words, the optimization routine did not find any departure from the steady-state measured values that would increase the ascendency.

A little reflection upon Fontaine's ingenious attempt to "test the hypotheses" shows that it is very difficult to unequivocally reject any of the several proposed principles. First, the reason ascendency met Fontaine's criterion to perfection can be traced to the fact that his perturbations always drove the system slightly away from the original steady-state. As can be inferred from (6.33) and (6.34), nearby to any steady-state, the imbalanced network configurations are likely to possess smaller ascendencies than that of the unperturbed state (by virtue of their having larger values of V in [6.34]). Hence, his algorithm was never able to find an initial direction away from the balanced starting point that would lead to a larger ascendency. Any arbitrary function that was locally maximal at steady-state would have likewise left the measured starting parameters unchanged.

More interesting would have been the implications had ascendency miserably failed Fontaine's test; that is, if increasing the ascendency had radically changed the starting configuration and parameters. Would that have been grounds to reject the principle of optimal ascendency? Certainly not! In any real system ascendency is kept from achieving its theoretical upper limit by various natural constraints. Whether those constraints are adequately represented in the mathematical formulation of the optimization algorithm is always open to question. If the objective function had radically altered the starting parameters, it could always have been argued that "hidden" natural constraints had not been properly expressed.

In short, optimal ascendency and related extremal principles are difficult to falsify. A negative result from any test either throws question on the adequacy of the test, or it forces a reformulation of the maxim under scrutiny. Fontaine was not testing hypotheses; he was evaluating the relative adequacy of various descriptions.

Even though it may be difficult to disprove any of the several proposed principles of development, it still should be possible to exercise some discrimination among them. After all, the geocentric theory of the solar system was never decisively disproven. One could always have preserved the quantitative accuracy of the Ptolemaic description by the addition of successive epi-cycles. Nevertheless, the earth-centered view of the cosmos was eventually abandoned with the advent of the more cogent Copernican theory. Inevitably, some of the tentative principles discussed here will likewise fall into disuse; however, the reasons for their rejection will probably not include their failing a single test, but will more likely involve a subjective consensus as to which quantitative narratives are more appropriate (see also Section 7.7).

Once one allows that influence might proceed down the hierarchy, it becomes necessary to reject the extremist notion that science should be confined strictly to the realm of testable hypotheses. That is not to say that most scientific endeavors may not proceed in the reductionistic vein of searching among falsifiable hypotheses. In cases where reductionism is appropriate, it is quite easy to imagine discrete phenomena at the lower levels as the cause for happenings at higher scales. In effect, one identifies a mapping from various events at one level focused on an event at a higher level. Each arrow in the mapping should be capable of falsification; and science, in the positivist's view, boils down to devising tests to winnow out the false arrows.

But when events at a larger scale affect those below, one does not simply reverse the direction of the arrows on the causal mapping. True, influence at a higher level is exhibited diffusely on the levels below, but effects are transmitted down the scale less by direct action (arrows) and more in the form of constraints on ensembles of smaller phenomena (Allen and Starr, 1982). To describe those constraints, as done by the laws of thermodynamics, is the crux of the phenomenological task. Insofar as any abstraction is perforce an approximation to reality, any description of constraints will remain, to some degree, inadequate. In the phenomenological realm, one is concerned less with binary situations of true and false, than with degrees of adequacy of description. It remains to be discussed how optimal ascendency might be further generalized to become a more adequate definition.

6.12 Summary

Once the assumption is made that an entity's organization can be sufficiently described by the structure of its transformations (flows), mathematical definitions for growth and development arise quite naturally. Growth is seen to be an increase in total system throughput; and development is a rise in the average mutual information of the network flow structure. Scaling the development by the throughput yields a quantity called the network ascendency, which reflects the ability of a system to prevail against other system configurations, real or putative. The ascendencies of systems with autonomous development are dominated by terms involving the internal (cyclical) transfers. The aggregate of these terms is called the internal ascendency. Increases in internal ascendency are subject to thermodynamic, hierarchical, and environmental constraints.

The principle of optimal ascendency synthesizes most of the attributes of developing ecosystems that were summarized by E.P. Odum. Ascendency is abetted by specialization, internalization, and cycling (especially recycling via the nonliving, more refractory components). Implicit in the increase of full ascendency is a tendency towards the steady-state balance of compartmental flows. As networks mature, the dominant factors in the rising ascendency switch from the growth of T (r selection) to the decrease in overhead Φ (K selection). The controversy over diversity and stability appears to be poorly drawn. A more appropriate concern is the inherent antagonism between A and R (the flow redundancy). The balance point between these two properties is determined by the rigor and unpredictability of the environment. In harsher environments, a higher fraction of the development capacity is encumbered by the redundancy.

When the medium of interest is energy, optimal ascendency translates into maximal work. Because a generalized work function always may be written as the sum of products of conjugate flows and forces, it becomes possible to define a formal "pseudo-force" conjugate to each ecological flow. These pseudo-forces, or selection pressures, are holistic in that their magnitudes depend upon the configuration of the whole ecosystem. The contribution of each system compartment to the overall ascendency is always positive, but this component should be properly compared to the gain in ascendency that would be possible if the resources used by the given species were otherwise utilized by the remaining members of the ecosystem. Selection is likely to act more strongly upon the dominant compartments of the network, thus allowing a large number of small compartments to persist at relatively little cost to the overall development capacity.

Optimal ascendency is intended to reconcile ostensible contradictions among earlier descriptions of growth and development. It is inappropriate to compare competing phenomenological descriptions for the purpose of deciding unequivocal acceptance or rejection. Just as a phenomenological statement cannot be completely verified, neither can it be entirely falsified. Some phenomenological principles become preferred over others due primarily to their being more adequate descriptions of events. The positivist attitude of outright acceptance or rejection of hypotheses is not without its attraction, but it is best confined to the reductionistic side of science.

7
Extensions

"Just as the essence of food cannot be conveyed in calories;
the essence of life will never be captured by even the greatest
formulas."

Alexander Solzhenitsyn
The First Circle

7.1 The Incomplete Picture

Just as it is impossible to completely falsify a phenomenological "principle," the obverse implies that any such description must always remain incomplete. Even the laws of thermodynamics were altered by the extension of observation into other spatial and temporal domains. The first law needed to be amended to account for relativistic considerations; and the second law, macroscopic as it is, is continually being contravened by chance occurrences at the molecular level.

One need not search very long to find the limitations of increasing ascendency as a descriptor of growth and development. A number of assumptions were made (not all explicitly) in formulating the principle. For example, the compartments were considered to be distributed homogeneously in space. Scarce mention has been made of temporal variations. Real systems circulate more than a single medium. How do events at one hierarchical level influence phenomena at higher and lower scales? These limitations and questions need to be addressed.

7.2 Spatial Heterogeneity

There is really no conceptual difficulty in extending ascendency to include spatial heterogeneities. In fact, spatial distinctions have been implicit in most of the networks considered thus far, and have been briefly considered in Chapters 3 and 6. In Figure 4.10, for instance, the bulk of the zooplankton are spatially separated from the benthic invertebrates—so too with the microphytes and macrophytes. Habitat (in the sense of location) is a prominent factor in identifying the compartments of an ecosystem.

However, it often happens that taxa occur with uneven densities over a wide variety of locations. In such cases, it is perfectly acceptable to define separate compartments for every taxon in each location where it occurs.

One then imagines an ecosystem of n taxa distributed over m spatial segments. The corresponding network would consist of $m \times n$ compartments.

It is usually impractical to realize such a representation. The dimension of a compound network climbs rapidly, and the work required to quantify such systems increases even faster. Depending upon the constraints, the number of flows to be estimated could rise exponentially; and it is rare that anyone treats networks with more than 50 compartments. Usually, an investigator opts to aggregate compartments so as to limit their number to 20 or fewer.

Example 7.1

Figure 7.1 depicts a hypothetical network of flows of carbon (g C m^{-2} y^{-1}) among the trophic levels of an open ocean ecosystem. The unprimed designations P, H, and C refer to primary producers, herbivores, and carnivores in the euphotic zone. The primed labels refer to the corresponding trophic levels as they exist in the aphotic zone. The arrows between P and P' and C and C' represent spatial movement of biota. In the case of the primary producers, the downward sinking (100 g C m^{-2} y^{-1}) is much greater than the turbulent resuspension. The reciprocal flows in the carnivore compartment represent diurnal vertical migrations in the water column. The C and C' compartments do not balance separately. The calculated ascendency for the system is 1718 g C bits m^{-2} y^{-1}.

The reader probably senses that by aggregating compartments, one loses information about the system. This loss appears explicitly as a drop in the mutual information factor of the ascendency. Hirata and Ulanowicz (1984) show that aggregating any two or more compartments can never lead to an increase in the average mutual information of the network structure. The drop in system ascendency is often even greater. This is due to the fact that intercompartmental flow between two nodes becomes

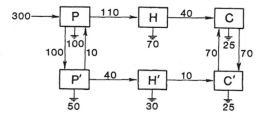

FIGURE 7.1. A hypothetical network of carbon flows (g C m^{-2}y^{-1}) among the trophic levels of an open ocean ecosystem. P, H, and C refer to primary producers, herbivores, and carnivores as they occur in the euphotic zone. The primed labels represent their counterparts in the deeper aphotic zone.

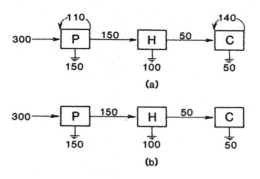

FIGURE 7.2. Depictions of vertical (spatial) aggregations of the flow network shown in Figure 7.1 (a) Maintaining account of the vertical migrations. (b) Ignoring spatial migrations. All flows are in g C m^{-2}y^{-1}.

intracompartmental cycling after the components are merged. If one has already quantified the highly resolved flow network, proper bookkeeping will define new intracompartmental cycles so that the total system throughput remains constant. However, one usually aggregates the nodes mentally before measuring flows; hence, the intracompartmental cycles are very often ignored. As a result, the measured total throughput is almost always an underestimate of the ensemble of processes that are occurring. Both factors in the ascendency may diminish because of aggregation, and the product almost always drops.

Example 7.2

If one ignores the demarcation of the euphotic zone in the ecosystem of Figure 7.1, the consequent spatially averaged chain appears in Figure 7.2a, where the migrations now appear as internal loops. The total system throughput of the system remains unchanged, but the information lost through the process of spatial aggregation decreases the ascendency by nearly 39% to 1048 g C bits m^{-2} y^{-1}.

However, should one be unaware of the internal circulations, the measured chain would then appear as in Figure 7.2b. By neglecting the migratory loops, the total system throughput falls by 24% from 1050 g C m^{-2} y^{-1} to 800 g C m^{-2} y^{-1}. The ascendency without loops is 974 g C bits m^{-2} y^{-1}, which is a drop of 43% from the original network in Figure 7.1.

If spatial aggregation serves to depress the ascendency, then, conversely, spatial diversification affords mechanisms by which to increase ascendency. Taxa that add little to the overall ascendency in a spatially homogeneous environment (i.e., are poor competitors in such a milieu) might contribute proportionately much more as specialists in a given spatial "niche." Separation into niches augments ascendency by increasing

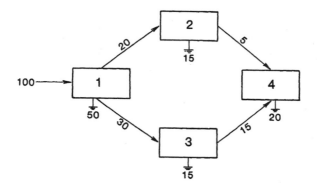

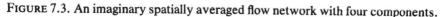

FIGURE 7.3. An imaginary spatially averaged flow network with four components.

both the number of compartments and the degree of specialization of each compartment. The evolution of spatial structure in an ecosystem thus serves to increase the network ascendency.

Example 7.3

A spatially averaged flow network appears in Figure 7.3. Compartment 3 has an efficiency of 50%; whereas compartment 2 is more dissipative, passing only 25% of its throughput on to compartment 4. The total throughput of the system is 270 flow units and the ascendency is 337 flow bits.

Despite its relative inefficiency, species 2 persists in the network. Further investigation reveals that there are two distinct spatial domains, one in which flow from 1 to 4 is via 2 and another where it is through 3, as shown in Figure 7.4. The total ascendency of this disjoint six compartment network is 403 flow bits. If the less efficient species 2 did not persist in its refuge, only the lower three element chain in Figure 7.4 would remain, with an ascendency of 201 flow bits. Hence, the hypothetical system in Figure 7.3 did not proceed to a configuration of greatest possible ascendency (all flow through 3) because of a "hidden" spatial con-

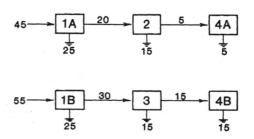

FIGURE 7.4. Resolution of the network in Figure 7.3 into two spatial domains, A and B. No communication between the zones is envisioned.

straint (the existence of a refuge) permitting species 2 to persist within its niche.

That ascendency may be defined in a spatially segmented realm prompts the speculation that the principle of optimal ascendency might describe autonomous behavior as it is observed in many dissipative physical systems. Development in meteorological systems comes immediately to mind, especially the growth and development of hurricanes. One can imagine a hurricane within a gridwork of many spatial cells, with one or a few media (e.g., water, air, and energy) circulating through each cell. Advection is cast as flow among the spatial elements. The ascendency of the consequent network should increase as the hurricane grows in intensity. The application of variational principles to meteorology is not entirely new. Paltridge (1975) used the principle of minimal entropy production to predict fluctuations in global climate patterns resulting from changes in insolation or land use.

7.3 Temporal Dynamics

While spatial variation poses no methodological difficulties to calculating the ascendency, it is slightly more difficult to accommodate temporal variation. It should be remembered that ascendency is defined in terms of information theory, which in turn rests upon probabilities. To treat spatial change required only redefining the categories (compartments), which does not affect the nature of the underlying probabilities. But to treat dynamic systems requires that one consider time-varying probabilities.

First, it should be noted that flow imbalance around a node poses no problem to calculating the ascendency. Equation (6.9) is valid even when the total inputs and outputs for each node are not equal (i.e., $Q_i \neq Q_i'$). Furthermore, it was shown using (6.33) how optimization of ascendency implicitly leads to a balance of flows.

However, being able to calculate an ascendency for an unbalanced network does not infer that the non steady-state dynamics are implicit in the optimization of A. To build temporal dynamics into ascendency, it becomes necessary to add another dimension to the description—that of time. All of the flow matrices discussed until now have been two-dimensional. That is, T_{ij} represents the flow of medium from i to j as accumulated over a suitable interval of time. By "suitable," it is meant that the interval should be long with respect to the fastest, macroscopic environmental changes and should preferably span the generation time of the longer-lived components. The interval also should be small with respect to the time over which system development takes place, so that the ascendency may be regarded as a function of time and its evolution may be charted.

Next, the assumption is made that the chosen interval can be divided into q equal subintervals, each of duration Δt. The subinterval from $t_o + (k - 1) \Delta t$ until $t_o + k \Delta t$ is then denoted by t_k; and, following previous notation, T_{ijk} will denote the flow of medium from compartment i to component j during the interval t_k. The number of subintervals q should be as large as necessary to characterize the fastest significant changes.

The accumulated flows, (T_{ij}) that characterize the averages can be retrieved from T_{ijk} by summation over the index k, that is,

$$T_{ij} = \sum_{k=1}^{q} T_{ijk}. \tag{7.1}$$

When T_{ij} is resolved into T_{ijk}, it describes a network flow trajectory in time. It now becomes possible to estimate the joint probability that out of all possible flow events occurring during the observation period, both a quantum of medium leaves a_j and a quantum of medium enters b_i during interval t_k, $p(a_j, b_i, t_k)$, as

$$p(a_j, b_i, t_k) \sim T_{jik}/T, \tag{7.2}$$

where $T = \sum_{i=1}^{n+2} \sum_{j=1}^{n+2} \sum_{k=1}^{q} T_{jik}$ (see 6.4). Once the three-way distribution has been estimated, it is easy to calculate the two-way joint probabilities, $p(a_j, b_i)$, $p(a_j, t_k)$ and $p(b_i, t_k)$ as well as the univariate probabilities $p(a_j)$, $p(b_i)$, and $p(t_k)$ by appropriate summations over the indices i, j, and k of $p(a_j, b_i, t_k)$.

Now the average mutual information (5.13) and, consequently, the ascendency (6.9) quantify the average information about the destination b_i of a quantum of flow that originates at a_j, assuming that all flows are averaged over time. However, if the flows vary with time, and furthermore if one knows the interval during which the transition took place, one should have more knowledge about the exact destination; that is, one has more information about the system.

McGill (1954) and Abramson (1963) discuss mutual information in multidimensional systems, and note that the appropriate generalization of (6.1) is:

$A(b; a, t)$

$$= K \sum_{i=0}^{n+2} \sum_{j=0}^{n+2} \sum_{k=1}^{q} p(a_j, b_i, t_k) \log [p(a_j, b_i, t_k)/p(b_i)p(a_j, t_k)], \tag{7.3a}$$

or, in terms of the flows T_{jik} (setting $K = T$),

$$A_t = \sum_{i=0}^{n+2} \sum_{j=0}^{n+2} \sum_{k=1}^{q} T_{jik} \log (TT_{jik}/T_i T_{jk}), \tag{7.3b}$$

where A_t has been used to denote the ascendency for a system with temporal variation. Knowing the temporal variation of the flows always requires more information (or delineates more structure, depending upon

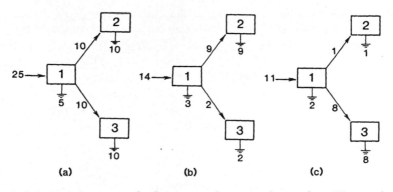

FIGURE 7.5. (a) A fragment of a flow network averaged over time. The resolution of (a) into two successive temporal subintervals is shown in (b) and (c).

one's viewpoint), but the temporal mutual information remains bounded by the averaged development capacity C; that is,

$$C \geq A_t \geq A \geq 0. \tag{7.4}$$

Inequality (7.4) implies that the average overhead $(C - A)$ will always equal or exceed the overhead of the time-varying system. Thus, it may be said that some of the average overhead has been "explained" by knowledge of the temporal variation of the system flows.

Example 7.4

The averaged flow values of a system fragment are depicted in Figure 7.5a. Equal amounts flow from compartment 1 to compartments 2 and 3, and there is ambiguity about where the flow leaving 1 will go. ($A = 92.4$, $C = 130.4$, and $\Phi = 38.0$ flow bits.)

However, when the flows are measured during two subintervals of time, the configurations in Figures 7.5b and 7.5c result. During the first interval, there is a higher probability that flow will proceed to unit 2. Flow to 3 is favored during the latter interval. The new knowledge about temporal variation in the system is reflected in a value for A_t of 100.4 flow bits. Eight flow bits of a priori overhead may now be attributed to temporal structure.

In Example 7.4, the averaged flows were considered to be an approximation of the more detailed temporal scheme as depicted in Figures 7.5b and 7.5c. As a result, the averaged network underestimated the real system ascendency. However, the example could also be interpreted as a system that was initially at steady-state and thereafter responded to variations in its environment through internal temporal changes. Of course, such responses would not be altogether arbitrary in that the temporal average should approximate a balanced network (see Section 6.4). The situation is not unlike that in other branches of non-linear dynamics, where the balanced steady-state is never achieved at any instant, but the system continues to oscillate in either periodic or aperiodic fashion around a virtual or "climax" configuration.

Insofar as the external environment exhibits some degree of regularity, systems should respond to optimize their ascendency under the prevailing conditions. Such adaptability is discussed at great length by Conrad (1983) and is implicit in the optimization of A_t. Even when environmental changes appear stochastic, they usually retain some degree of temporal structure, which may be characterized by the Fourier spectra of the environmental variables. A Fourier analysis of an environmental time series (e.g., of velocity, temperature, salinity, etc.) calculates how much of the overall change can be characterized as occurring within a given frequency range. Real environmental noise is never purely stochastic (i.e., a flat frequency spectrum, or white noise [James Hill IV, personal communication]), but it usually exhibits peaks, slopes, and other discernible structural features.

Biotic systems respond to the environment in ways that also can be characterized by Fourier spectra (Platt and Denman, 1975; Powell et al., 1975). Insofar as the responses serve to increase A_t, they may be labelled adaptability. Intuition suggests that adaptation would be facilitated by mechanisms for memory within the system infrastructure. Interpreting the relationships between input and response spectra in terms of the maximization of A_t should result in an intriguing narrative of ecosystem function.

7.4 Multiple Media

Discussion until now has focused on a single medium. The presumption has been that the effects of other media are implicit in the network structure of a single key medium. In ecosystems, the key medium most often is taken to be energy or energy's most common carrier, carbon. In economic systems, it is usually currency.

However, real living systems are manifestly more than single networks. In ecology, it is also common to measure fluxes of nitrogen, phosphorus, silicon, and many other trace elements, each of which defines a separate network. In economics, one quantifies multiple commodity networks. Is it possible to make explicit the effects these parallel circulations have upon the flow network of the key medium?

One obvious approach would be to treat multiple media as an extra dimension in the same manner that time was handled in the last section. If it were possible to convert all media into common units, then one could define a three-dimensional array (T_{ijl}) to represent the flow of medium l

from compartment i to compartment j. The definition of multimedia ascendency would parallel (7.2ff.), with s_l replacing t_k and the number of separate media (i.e., m) replacing the number of time increments (q). Growth and development in the multimedia system would then appear as an increase in

$$A_s = \sum_{i=0}^{n+2} \sum_{j=0}^{n+2} \sum_{l=1}^{m} T_{jil} \log(TT_{jil}/T_i T_{jl}), \tag{7.5}$$

where T is the total flow expressed in the common units.

The problem at hand is to identify appropriate conversion and weighting factors for each of the separate media. That is, one wishes to define ψ_l such that

$$T_{ijl} = \psi_l T_{ij}, \tag{7.6}$$

where ψ_l is the weighting factor for medium l, which transforms all the flows into equivalent units. How best to establish the ψ_l is still unsolved, but one expects that clues to the solution exist in the economic theory of pricing (Amir, 1979; Constanza and Neil, 1984).

Example 7.5

Figure 7.6a depicts a network of flows of medium A and Figure 7.6b, a parallel network of medium B. Both media are expressed in common units such as mass. The ascendency for the combined network is 329 flow bits, whereas if the two media are considered distinct, A_s becomes 334 flow bits. Hence, only 5 of the original 190 flow bits of overhead are "explained" by the flow network of medium B.

In Figure 7.6a, it is noted that compartment 2 is far more efficient than compartment 3. If one were to optimize the ascendency of this configuration without regard to medium B, the pathway 1-2-4 would eventually dominate the network, and species 3 would be pressed towards extinction, as in Figure 7.7a. However, if compartment 4 depends upon medium B for survival, then the extinction of compartment 3 would collapse the system to the degenerate form in Figure 7.7b. To go from the configuration in Figure 7.7b to that in Figure 7.6a requires the "expenditure" of 67.4 flowbits of overhead to achieve a gain of 40.2 flow bits of ascendency. (N.B., $A = 299.1$ flow bits for the network in Figure 7.6a alone.) In light of these differences, it appears that significantly more than 5 of the 190 flow bits of overhead are devoted to maintaining the flow of B to species 4. This last comparison underscores the need for a pricing scheme appropriate to the multimedia problem.

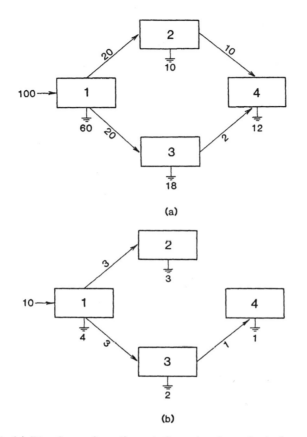

FIGURE 7.6. (a) The flow of medium *A* through a hypothetical network of four species. (b) The flow of medium *B* through the same network.

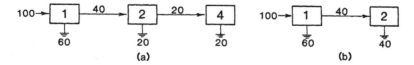

FIGURE 7.7. (a) The result of optimizing the ascendency of the network in Figure 7.6a without regard to the existence of medium *B*; but this action neglects the dependency of 4 upon 3 (as shown in Figure 7.6b), thereby also extinguishing 4. The final result is schematized in (b).

7.5 Overall Heterogeneity

In light of the last two sections, it becomes apparent that the flows in a spatially heterogeneous environment could have been regarded as the elements of a four-dimensional array (T_{irjs}) representing the flow of medium from species i in cell r to species j in cell s. Carrying this line of reasoning two steps further, it is possible to regard the flows in a real system with simultaneous heterogenity in space, time, and matter as the components of a six-dimensional array (T_{irjskl}), where this arbitrary element refers to the magnitude of flow of medium l in species i at location r to species j at locus s during the time interval k. The appropriate ascendency, call it A_u, would be calculated in a manner similar to that yielding (7.3a) or (7.5), as

$$A_u = \sum_{i=0}^{n+2} \sum_{j=0}^{n+2} \sum_{r=1}^{p} \sum_{s=1}^{p} \sum_{k=1}^{q} \sum_{l=1}^{m} T_{irjskl} \log (TT_{irjskl}/T_i T_{rjskl}). \tag{7.7}$$

A_u would take on a value between C and A.

In Section 7.2, it was argued that a system initially uniform in space could increase in ascendency through the creation of spatial structure. In Section 7.3, temporal variation was likewise seen to be an avenue by which A_u could increase. Finally, one may infer from Section 7.4 that diversification of material constituents also could facilitate a rise in ascendency. One is left with the impression that the continual increase in A_u reflects a progression toward more "baroque" forms, which is a tendency commonly observed. In Chapter 6, however, increasing ascendency appeared to simplify the network by driving inefficient compartments and pathways to extinction. Paradoxically, increasing ascendency seems to be both a drive towards simpler form and a tendency towards the more complex configurations (Emery, personal communication; see also Prigogine, 1980).

Confusion on this point may be avoided by focusing on increasing ascendency as an indication that selection is at work. When the hypothetical environment is simplistically constrained, and the number of compartments is not allowed to increase, the selection mirrored by increasing ascendency cannot but simplify the given network. However, the natural world is never so rigidly constrained. Spatial, temporal, and material variations in exogenous factors are complicated to say the least, and the potential combinations of environments and system states surpass the imagination. (See Elsasser, 1981 for a revealing discussion on immensity.) From among the virtual infinitude of possible future states, the system "selects" a unique trajectory; and that selection, on the average, is in the direction of increasing system ascendency. Thus, the observer of natural systems, subject as they are to loosely constrained environments, perceives developing systems as becoming ever more complicated. However, even complicated real structures appear amazingly simple against a background of such immense potential complexity. Increasing ascendency, therefore, consistently appears as the reflection of ruthless selection.

7.6 Aggregation

It is obviously impractical to measure the full array of flows, T_{irjskl}. For this reason, one usually resorts to some form of averaging over space, time, or substance. Still another way to reduce the dimensionality of the system is to combine components; that is, aggregate the nodes. Of course, lumping species degrades the description, so the question naturally arises how to minimize the impact of grouping nodes.

Halfon (1979) refers to aggregation as the identification problem and devotes a substantial section of his book, *Theoretical Systems Ecology*, to the subject. In ecology, one is usually forced to aggregate prior to measurement; otherwise quantifying the network becomes infeasible. It is rarely necessary to condense ecological flow data after the fact, although such necessity often arises in economics. Nonetheless, studying a posteriori aggregation should shed light on how to best identify system compartments at the outset.

If one accepts the principle of optimal ascendency as a desirable description of system growth and development, it follows that one wishes to aggregate in a manner that distorts that description the least. That is, the act of merging components should leave the condensed network with the highest possible ascendency. Hirata and Ulanowicz (1984) have proven that it is impossible to increase the ascendency by consolidating nodes. The problem, therefore, is to find an aggregation scheme that decreases network ascendency the least.

A desirable constraint upon any aggregation scheme is that the total system throughput remains unchanged by the condensation (see also Ulanowicz and Kemp, 1979). Hence, changes in ascendency under any aggregation that keeps throughput constant must be due entirely to variations in the network mutual information. In other words, one seeks to combine compartments in such a way as to minimize the loss of information about the network. This last sentence describes a goal that should have broad intuitive appeal.

There seem to be few choices as to how to minimize the loss of network information. The only algorithm certain to identify the optimal aggregation appears to be a brute force search over all possible combinations. As discussed in Section 4.4, such combinatorial searches are generally infeasible when the number of network nodes exceeds about $n = 10$.

The only feasible strategy to treat networks where $n > 10$ bears analogy to stepwise regression techniques. Starting with n nodes, one searches all pairwise combinations of compartments for the one that least reduces the ascendency. The optimal pair is coalesced, and the search is repeated on the $n - 1$ dimensional network. The search is iterated until the system has been aggregated into m compartments. There is no guarantee that the network resulting from this scheme will have the greatest possible ascen-

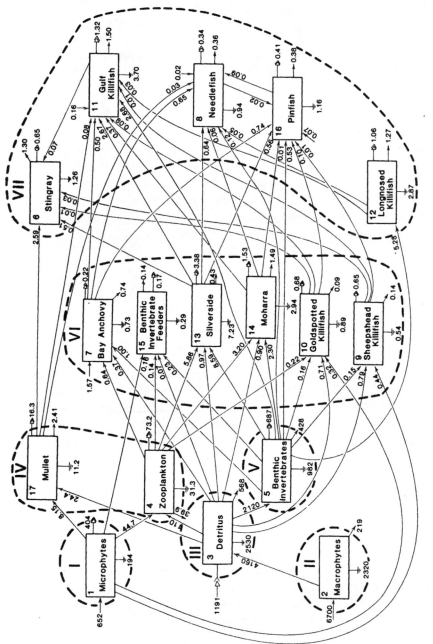

FIGURE 7.8. The aggregation of the Crystal River flow network (Figure 4.10), which yields the highest ascendency when condensed in pairwise fashion from 17 to 7 compartments. See Figure 4.10 legend for further details.

dency, but in this author's experience, most results have been very close to the optimal. The number of calculations necessary for the pairwise coalescence scheme increases only as n^2, making it feasible to apply this algorithm to networks with 20 or even 50 compartments.

Example 7.6

The aggregation of the Crystal River ecosystem network (see Figure 4.10), which yields the highest ascendency when condensed in pairwise fashion from 17 into 7 compartments, is shown in Figure 7.8. Flows in the simplified seven compartment network are depicted in Figure 7.9. The ascendency fell only minutely, from 28,337 to 28,287 mg C bits m^{-2} d^{-1}.

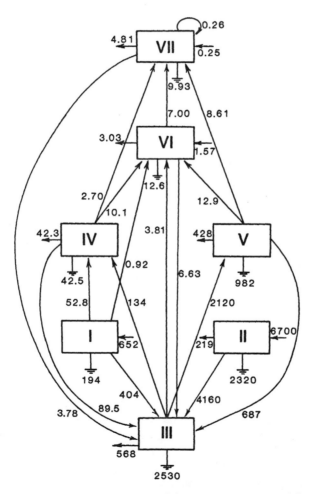

FIGURE 7.9. The pictorial consolidation of the compartments and flows in Figure 7.8 into the optimal seven node network. Flows are in mg C m^{-2} d^{-1}.

What is remarkable about these results is that Figure 4.10, as amended in Figure 7.8, was prepared for another publication several months before the topic of aggregation by optimal ascendency was ever addressed. The placement of the boxes on the diagram was the result of much trial and error effort to make the schematic as uncomplicated as possible. To see intuitive guesswork confirmed, as it were, by blind calculation was both startling and exciting. However, such agreement should not have been altogether unexpected when one remembers that information theory was *designed* to quantify the intuitive notion of information, while ascendency was *intended* to capture the common meaning of organization in mathematical form.

This example also illustrates that relatively little interaction (in terms of transfers) exists within trophic levels (e.g., compartments IV, VI, and VII) (see "congeneric homotaxis" in Hill and Wiegert, 1980). This insight serves to counter some of the criticism levelled against the idea of ecosystem organization. Simberloff (1980), for example, relies on the data of Gleason (1926) concerning the "individualistic" behavior among the members of a plant community as evidence that the notion of organized ecosystems is ill-considered. However, organization is seen to exist over the total ensemble of transfers, and the point made here is that the least likely place to discover it is where Gleason confined his search—among the members of the same trophic level.

7.7 Ascertaining Configurations of Optimal Ascendency

Heretofore, optimal ascendency has been referred to in the abstract, or at best illustrated in analytical fashion for very simple networks. However, if the principle of optimal ascendency is to achieve any widespread application, it becomes necessary to create an algorithm to optimize the ascendency of an arbitrary network possessing a moderate number (i.e., 20 to 50) of compartments.

The objective function (the ascendency) in the proposed optimization is nonlinear, and nonlinear optimization problems are notoriously difficult. This is particularly true in the present case, since the mathematical function to be maximized has neither of the "regularizing" qualitative (or smooth) properties (concavity, convexity) that justify most solution techniques. Furthermore, even the mass balance constraints are nonlinear without the rather restrictive assumption that dissipations and exports be fixed fractions of their corresponding throughputs (T_i's). To deal with these difficulties, Cheung and Goldman (Cheung, 1985a, 1985b) have developed an algorithm that proceeds by successive linear approximations, adapting generalized network techniques (Kennington and Helgason, 1980) to solve at each stage the resulting "minimum-cost flow in a network with gains" problem.

Full technical details of the optimal ascendency algorithm are beyond the scope of this text. The important matter here is that it is possible to start with any observed network of moderate size and complexity, and to ascertain what the most "efficient" configuration would be if all hierarchical and environmental constraints were absent. The differences between the observed and the optimal configurations should highlight where in the network pressures for growth are most likely to occur. Conversely, describing those hierarchical and environmental constraints that keep the system from attaining its local ascendency maximum appears to be a worthwhile exercise in descriptive community ecology. Such constraints might then be incorporated in the numerical optimization procedure sketched above.

The opportunities afforded by the ability to optimize network ascendency are manifold. There are plans to simulate the *process* of succession using increasing ascendency as a selection criterion. The intention is to commence the simulation with an arbitrary, simple network and to intermittently subject the system to an "environment" of stochastic perturbations. New compartments and flows of miniscule magnitude could be introduced in random fashion, and existing nodes and flows could be extinguished according to probabilities of extinction that are related in some inverse way to their respective throughputs. Between environmental events, the existing network would be allowed to progress towards its local ascendency maximum. The parameters specifying the stochastic environment could be adjusted to look at the effects that various environmental conditions might have on the configuration of the "climax" community.

The successive linearization technique used to optimize the network ascendency may also be applied to optimize other network attributes, such as total systems throughput (T), network redundancy (R), or dissipation (S). Therefore, it becomes feasible to change objective functions and repeat the simulated network succession in an equivalent stochastic environment. In effect, optimization runs on T, R, and S would be caricatures of development under the proposed principles of Lotka, E.P. Odum, and Prigogine, respectively (see Section 6.11). By comparing the qualitative features of the different climax communities created by the four objective functions (ascendency included), it may be possible to render some preliminary judgments concerning the relative adequacy of the different descriptions of growth and development (see Fontaine, 1981).

7.8 Other Applications—Economics and Ontogeny

Throughout this discourse, there have been scattered references to autonomous development in fields of study other than ecology. Odum (1977) takes the position that ecology is emerging as an integrator for the disciplines of economics, sociology, and political science. Whether economics is merely another form of ecology is certainly questionable (Daly, 1968), but what is truly integrative of these disciplines is the common trait of growth and development, which admits a unified description.

Unfortunately, few individuals are truly fluent in more than one of these separate fields of inquiry. This author does not pretend to be an expert on several fronts, but nevertheless cannot resist the temptation to mention how ascendency might be applicable to a layman's view of events in economics and ontogeny. It is to be hoped that out of curiosity, developmental biologists and economists will critically pursue the following casual analogies.

"Economics" and "ecology" share the same etymological root—a fact suggesting that those who coined the latter term probably perceived dynamics that were common to both fields. Growing interest in the interactions of natural and human systems has increased the exchange of ideas between the two academic fields (e.g., Ayres and Kneese, 1969; Isard et al., 1972; Victor, 1972). The most notable overlap has been the development of quantitative models addressing various exchanges between the economic and natural domains (e.g., Isard, 1968; Odum, 1971). (As an interesting aside, it should be noted that relatively fewer economists are fettered by reductionist disdain for "macroeconomics," or for the study of the influence of global conditions upon microdynamics [see Lange, 1971].)

As discussed in Chapter 3, at least one economic tool (input-output analysis) has been introduced into ecology, with the result that an increasing number of ecologists are beginning to think of ecosystems as networks of processes. However, ecologists as a whole still do not appreciate how their networks differ from those of the economist.

To begin with, in economics, one must deal with flows in both directions: commodities flowing one way and currency in the opposite sense (Odum, 1971). However, currency networks are not merely the photographic negatives of physical flows (Boulding, 1982). For example, there is no strict microeconomic counterpart to respiration. While the physical or biological scientists might expect capital depreciation, handling fees, legal charges, and similar "dissipative" flows to be the economic analog of respiration in biophysical networks, they would be disconcerted to learn that the corresponding element in the currency network is occupied by consumer demand. Diffuse flows of currency from consumers becomes the major income for many corporations. New categories of inputs appear that correspond to value-added capital and labor costs—both of which occur without appreciable conjugate material fluxes. In the end, only the interindustry payments (the f_{ij}'s or g_{ij}'s) remain strictly analogous. In contrast to energy and material networks, the network of cash flows appears almost closed—even the "dissipative" flows result in money in someone's pocket.

Hence, the four-category scheme developed in Chapter 3 to classify energy and material flows may not best categorize currency flows. As long as the crucial distinction between inputs and outputs is maintained, the definitions of the key variables (ascendency, development capacity, and overhead) are not qualitatively affected by how the types of inputs or outputs are distinguished.

As to what constitutes dissipation at each node of an economic network, thermodynamically speaking, the issue is moot. The second law of thermodynamics describes events at the macroscopic level. What happens microscopically need not be irreversible (although Prigogine [1978] argues that microscopic irreversibility is probable). What matters in the end is that macroscopic overhead always remains non-negative. That is, not all of the development capacity of an economy can appear as size and organization, even if strict accounting is required for each economic sector. The non-negative redundancy ensures that the second law will be satisfied at the macroeconomic scale.

The gross output, or gross national product (GNP), is still generally viewed as the foremost index of macroeconomic performance. However, the GNP is a component of the more inclusive total system throughput, which in turn is modified by a structural factor in the ascendency to account for the coherence of the cash flows. Furthermore, the various components of the economic ascendency could serve to indicate how the network might be restructured to best augment the general wealth. Therefore, the ascendency of the cash transfer network has much to recommend it as an alternative index of economic vigor and efficiency.

Example 7.7

Figure 7.10 depicts a simplistic hypothetical economic community of four corporations. All flows are measured in millions of dollars per year. Ex-

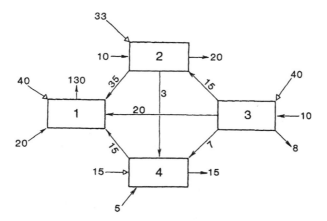

FIGURE 7.10. Diagram of economic transfers (in millions of dollars y^{-1}) among four hypothetical corporations. Exogenous inputs with regular arrowheads represent purchases of goods, whereas those inputs with triangular heads indicate nonmaterial services (e.g., labor, management, profits, taxes, etc.).

ternal inputs to the compartments are differentiated as being associated either with the purchases of goods (represented by the regular arrows) or with value-added inputs of nonmaterial services such as labor, management, profits, taxes, etc. (shown by the triangular arrowheads).

Corporation 1 is a producer of goods to meet final demands; otherwise known as a "means of final consumption." Corporations 2 and 4 produce intermediate goods, or "means of production," and corporation 3 deals in nonmaterial services (e.g., accounting, law, etc.) catering to the needs of the other three establishments. The gross community product of the ensemble is $183 million y^{-1}, and the total systems throughput (or global production) is $441 million y^{-1}. The development capacity of the group is $905.8 million bits y^{-1}, of which 42.5% appears as ascendency.

It is interesting to study the components of the ascendency. Table 7.1 lists the contributions in dollars $\times 10^6$ bits y^{-1}:

TABLE 7.1. The components of the ascendency (in $ $\times 10^6$ bits y^{-1}) of the hypothetical community in Figure 7.10 arrayed in matrix form.

	1	2	3	4	Value added	Imports	Final demand	Supply sums
1	0	0	0	0	0	0	176	176
2	36.2	0	0	−1.2	0	0	−3.7	31.3
3	8.8	17.8	0	7.3	0	0	−10.4	′23.6
4	11.4	0	0	0	0	0	5.3	16.7
Value added	3.4	32.0	58.5	11.8	0	0	0	106
Imports	11.8	7.6	9.7	3.5	0	0	0	32.7
Final demand	0	0	0	0	0	0	0	0
Demand sums	71.6	57.5	68.2	21.4	0	0	167	385.4

Regarding the negative entries in Table 7.1, it is probable that pressures for change will be focused on the sales from 2 to 4 and on the final demands for goods and services from 2 and 3. Neglecting any hidden constraints, the economic ascendency of the community (a measure of the overall efficiency) would be fostered by industry 2 focusing more of its sales on 1 and giving less attention to demands from either 4 or the external world, and by corporation 3 attending only to demands from within the community. Of course, constraints to these revised strategies probably do exist. For example, it is possible that the commodities involved in the proposed substitutions are not equivalent, making the suggested alterations unfeasible. (See "hidden constraints" in Section 7.4.)

Also interesting are the dual contributions (marginal sums in Table 7.1) that each corporation makes to the ascendency. For example, from the supply side, corporation 1 contributes $\$176 \times 10^6$ bits y^{-1} (46%) to the total ascendency. From the demand side, however, it contributes only 19% to the same total. This difference exists despite the balance of supply and demand for each component. It is no wonder, then, that supply and demand side economists argue over their sometimes conflicting perspectives on an economy. However, overall ascendency is symmetrical with respect to supply and demand; that is, $A(a; b) = A(b; a)$. It is an unbiased criterion with which to evaluate structural changes in an economy.

Before leaving economics, it is worth mentioning some casual observations on how developing economies appear to increase in ascendency. For example, the early American pioneers were faced with an incredible wealth of natural resources. The population rose precipitously (fueled largely by immigration) to exploit available natural inputs. The accompanying social climate was one of egalitarianism. Most pioneers were highly self-reliant generalists; there being one predominant economic element (subsistence farming) and relatively few other occupations (tinker, tailor, soldier, etc.). Development capacity was rapidly increasing, primarily because the throughput of the aggregated individual homesteads was growing. Comparatively little of this development capacity appeared as ascendency, however, due to the relative isolation of the components and to the diffuse nature of existing economic links. The finitude of space and resources inevitably served to slow down the rise in capacity; and with the wave of industrialization came increasing occupational specialization, which mandated greater interdependence. Competition and feedback, as in ecology, served to increase ascendency by accentuating the most efficient pathways (e.g., through mergers, bankruptcies, and the like).

In a laissez-faire economy, this increasing ascendency usually means a tendency towards monopoly—the survival of the most efficient. However, a disproportionately high ascendency or efficiency is likely to handicap the adaptability of the system to change. The existence of at least minor parallel pathways to compensate for the unexpected debility of a major supplier gives the economy a strength in reserve. Also in a monopolistic, unbridled economy there is a proclivity to cycle wealth among the "upper trophic levels" (members of the upper economic strata). The near absence of cycles among the higher trophic elements of natural systems suggests that such circumstances also lack persistence. It is not too surprising, therefore, that most mature economies have developed controls (e.g., antitrust legislation or social welfare programs) that in the short term slow the rise of economic ascendency, but otherwise lend homeostasis to the community and thereby allow it to take on higher ascendency in the long-term.

It is one thing to moderate a natural economy and totally another to attempt to design an economic structure. Centrally planned economies, in which there is deliberate control over the appearance of new centers of production, usually end up having more simplified networks of industrial exchanges. Of course, the fragility of such designs is at least as great as that of a natural community of monopolies. However, a still greater risk in attempting to mandate prices and production rates is that one must inevitably overlook, and often thwart, most of the feedback controls operating to increase the ascendency of the economic network. Autonomy is rather difficult to plan! It would seem preferable to allow some chance to operate at the microeconomic level to permit the natural development of vigorous feedback pathways.

The point of this little discourse has not been to overlay truism with platitude, but to indicate the possibility of pursuing economic narrative and debate using the vocabulary developed in this text. The advantage of addressing economics in these terms is that parallels with natural history immediately become evident, thereby multiplying the wealth of observation from which to draw economic conclusions and recommendations.

Ontogeny, the study of the development of individual organisms, must deal with certain factors that are either missing from or relatively insignificant to ecology and economics. Foremost among these factors is the very obvious goal-directedness of ontogeny. If one knows the origin of some seed material, it becomes possible to predict within reasonable bounds the pattern that organismal development will take. The regularity of this evolution and the correspondence of the final organism form to that of its progenitors gave rise to the abstract notion of genes. Brilliant progress has been achieved during the past 50 years in identifying a material locus for genes and in describing copious molecular mechanisms at work in the ontogenetic process.

Unfortunately, the elation accompanying these discoveries has engendered immoderate attitudes in some individuals. Those who continue to adhere to strict genetic determinism remain quite numerous and often influential. Many of the same persons who would deride an ecologist who espouses variational principles as a teleologist or vitalist seemingly have no compunction about ascribing extraordinary powers to molecules. Ribonucleic acids do not serve simply to encode information; they are said to actively *direct* the development of the organism. They even take on human-like characteristics, such as selfishness (Dawkins, 1976). The phenotypes, by contrast, merely mark time between the only events that count—the reshuffling of DNA molecules. Higher forms of life, and most certainly consciousness, are epiphenomenal—of secondary or lesser interest. DNA, it is implied, ultimately determines how a rocket will be constructed or a symphony composed. All power and directedness is considered to flow from the molecular (or below) *up* the hierarchy of phenomena.

Certainly, genes strongly influence the developing pattern of the organism. However, is such causality sufficient to understand the ontogenetic

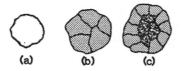

FIGURE 7.11. A pictorial representation of early blastular formation. (a) Growth begins as a single germ cell. (b) Initial division leaves all members with nearly equal access to supporting external medium, but after further growth, some cells are cut off from direct communication with the outer environment, resulting in two classes of cells, inner (dotted) and outer, as in (c).

process, not to mention changes that occur over several generations? Weiss (1969) argues rather forcefully against the sufficiency of molecular cause. As the components of the organism increase, there are "certain definite rules of order [that] apply to the dynamics of the *whole* system" (italics by Weiss). Waddington (1968) coined the term "canalization" to denote the effects of this unknown rule of ordering.

However, in concrete terms, how might the whole organism cooperate in guiding its own development? This enigma is addressed by a very simple, yet powerful example of Weiss (1969): "The model, diagrammed in Figure [7.11] starts from a single unit, let us say, a cell (a), assumed to be internally homogeneous, but in steady equilibrated interaction (e.g., exchange of substance and energy) with the ambient medium across its free surface. Let us then increase the number of such units, either by cell division or aggregation (b). At first all members of the group are equal and share in the free surface, hence continue to exist as before. But as their number increases, when a critical number is reached, a new situation suddenly emerges (c) which splits the mass of previously equivalent units into two radically disparate groups—an outer group, still in direct contact and exchange with the original medium, and an inner one, shut off from access to the old medium and encircled instead completely by its former equals. As the inner units (cells), in response to this drastic change of their conditions, switch into a different (metabolic) course, the formerly uniform group acquires a core-crust (inner-outer) differentiation. . . This inner-outer differentiation is one of the most elementary instances of a group pattern being definitely predictable as a whole, while at the same time the fates of its components are still indefinite, thus illustrating macrodeterminacy in the absence of microdeterminacy, as in Figure 7.11. Yet, the main point is not just this early indeterminacy of the members of the group, but the fact that their future fate is determined by their place within the dynamic configuration of the whole group and can be predicted and understood only in reference to the 'whole'."

Although Weiss does not define a macroscopic architectural principle, it is not overly presumptuous to suggest that blastular development is another example of increasing ascendency. Undifferentiated cells must meet all of their own needs; for example, transport of food source into the cell as well as manufacture of cell structure and metabolites. The blastula increases its ascendency by separating these roles. The outer cells can become more efficient at absorbing and transporting foodstuffs and expend less energy on the manufacture of metabolites. The interior cells in their turn can focus more of their effort upon anabolism and return some of their products to the outer cells.

As with ecological and economic development, early organism growth appears mainly as development capacity generated by increased throughput and compartmentalization. With subsequent specialization comes increased organism ascendency. In most forms of life, especially the higher phyla, the organism's development capacity eventually plateaus. Exactly why this happens is unclear, but Elder (1979) suggests it may be linked to an incompatibility between growth and memory. The stasis in C notwithstanding, A may nonetheless continue to rise at the expense of existing redundancy R. (This situation is perhaps best illustrated during the later development of the central nervous system of higher animals. Division of nerve cells in humans, for example, virtually ceases at about age 3. By such time, some of the nerve cells have been "prewired" or connected to other specific axons. However, many of the brain cells are equivocally connected to hundreds of other cells in their immediate vicinity. The pathway redundancy of the juvenile neural system is immense, and its development capacity is maximal. Early stimuli proceed through this maze in almost random manner. However, after repeated use, certain pathways are strongly reinforced, while others are inhibited. This neural articulation that accompanies learning increases network ascendency at the expense of redundancy.) Redundancy falls, not only in support of development, but also due to chance destruction of elements (injuries). The organism becomes progressively more vulnerable to disturbance (undergoes senescence), losing its "strength-in-reserve," and it eventually succumbs to some chance event.

If normal growth and development can be interpreted using information variables, might not pathologies admit a similar narrative? Cancer, for example, appears to be a breakdown in the normal course of development just described. There is a sudden retrogression to the growth of undifferentiated cells. It may be viewed as a breakdown in the whole organism regulatory process, and yet the key to understanding cancer is generally believed to lie at the level of subcellular events. Doubtless, the etiology of cancer is concerned with molecular processes; but just as an understanding of ontogeny is incomplete without recourse to whole level phenomena, so it is not necessary to be a devotee of "holistic medicine" in order to appreciate the need to direct some attention to describing whole organism regulation as it pertains to pathologies.

7.9 Summary

No phenomenological principle is a perfect description of nature, although the laws of thermodynamics appear adequate to describe an enormous range of catalogued phenomena. The narration of autonomous organization in systems has yet to achieve the generality of the two thermodynamic laws.

The barriers to generalizing optimal ascendency are more technical than conceptual. For example, spatial heterogeneities can be accommodated simply by reformulating the definition of a network compartment. A compartment may be taken as a given taxon occurring within a specific spatial subdomain. This redefinition has the disadvantage of sometimes dramatically increasing the network dimension, although spatial definition thereby becomes an explicit property of the expanded network. Carrying this artifice to an extreme, one may envision a network of spatial cells without taxa over which a single medium is coursing and dissipating. Optimizing the ascendency of such a system might provide a description of autonomous behavior as it occurs in physical systems, such as meteorological storms or galactic clusters.

To incorporate dynamic behavior into ascendency, it becomes necessary to employ concepts from probability and information theories as they pertain to higher dimensions. Inputs, outputs, and subintervals of elapsed time become the three dimensions of an expanded input-output array. The ascendency over the entire duration is defined as being proportional to the mutual information between the inputs and the joint occurrences of outputs and time intervals. The definition of ascendency for time-varying systems should allow one to interpret the results of adaptability theory and ecological spectral analysis as manifestations of the tendency of living systems towards increasing ascendency.

Dimensional ambiguities appear to forestall the definition of ascendency for systems wherein several media are flowing. If it were obvious how to interconvert media, the extension of the ascendency concept to multimedia networks could mimic the stratagem for incorporating temporal variations; that is, the several media could be treated as elements in a third dimension. Under such a formulation, many of the costs hidden in the flows of one medium could be interpreted as serving to facilitate the flows of other media.

Aggregations that minimize the loss of ascendency (information) accord very well with intuition, lending plausibility to ascendency as an apt characterization of organization.

Although ascendency was developed primarily in ecological terms, growth and development is common to other disciplines as well. Defining the common ground between development in economics and ecology, for example, may prove to be largely a lexical exercise. However, the history of ontogeny makes it difficult for many developmental biologists to allow the idea that a degree of autonomous regulation may occur at the level of the whole organism. Nonetheless, it is maintained that strict reductionism is insufficient to describe all causality in ontogeny.

7.10 Epilogue

Before ending this discourse, an attempt to clarify certain outstanding issues concerning ascendency seems to be in order. Probably the foremost objection that many will have to this thesis is that it is ostensibly rife with teleology; that is, some of the terminology has a distinct anthropomorphic flavor. If something is being optimized by a system, then the ensemble is approaching a "goal," and the temptation arises to look for a thinking creature able to identify the goal, make decisions, and effect changes (Straskraba, 1983). Some early investigators were even inclined to regard the ecosystem as a "superorganism" (Clements and Shelford, 1939).

The latter reference to superorganisms by Clements is often used against those seeking to measure attributes of whole ecosystems (see Simberloff, 1980). Clements' choice of terms was unfortunate, because the prefix "super" may refer either to scale or to ontological status. It is the latter meaning that Clements' detractors prefer, because it allows them to caricature his logic as something akin to the following syllogism:

> Fish have eyes.
> Humans have eyes.
> Therefore, fish are angels.

Although some ichthyologists believe this conclusion, it is an obvious nonsequitur. However, the faulty inference is hardly a reason for comparative physiologists to abandon the study of animal eyes. They are aware that the more appropriate conclusion is simply that both fish and humans have organs that extract information from light.

Similarly, by saying that whole ecosystems exhibit growth and development, it does not follow that ecological communities are necessarily equivalent to or somehow exceed the organizational or ontological status of an individual organism. Organisms and ecosystems merely share a common attribute. As an argument to dissuade ecologists from considering whole system properties, the taunts against superorganisms are gratuitous, at best.

Returning to more conventional anthropomorphisms, the anxieties associated with teleology seem most acute in the ontogenist, who is faced with the very real observation that developing organisms *do* exhibit goal-directed behavior. If one knows the origin of the germ material, the final form (goal) of the mature organism is known to within reasonable limits.

The dilemma of the developmental biologist is how to make sense of organism development without invoking anthropomorphic behavior at any hierarchical level.

The resolution echoed here is that ontogeny is an integration of influences both up and down the hierarchy. Material antecedents, to a large extent, define the final form (goal) of the organism, but the next step anywhere during the course of development is also influenced by the state of the entire organism. This whole system influence is labelled autonomous development and is quantified by an increasing network ascendency.

To see why this position is not necessarily teleological, it helps to consider ecological and economic systems, wherein neither a material template nor an end goal for development is at all apparent. Yet growth and development in these domains is fundamentally the same behavior exhibited by growing organisms. Of course, at any particular time, it is possible to argue that the ecosystem is approaching a specific goal (namely, a local maximum of ascendency). However, that apparent goal is a result of the present and past states of the system. Events in the near future can radically change the apparent goal. (There even exist physical systems subject to strict deterministic laws, but which approach no precise goal or regular pattern of behavior [see Ulanowicz, 1979 for review].)

An appropriate analogy of an indefinite goal is the parlor game made popular by the physicist John Wheeler. Several people are at a dinner party, and as predinner entertainment, the host suggests they play a game. One person is sent out of the room to return shortly and guess a word that has been chosen by the remaining group. While the guesser is out of the room, the host proposes that the group *not* select a word. Instead, the first respondent may arbitrarily answer "yes" or "no" to the first query. The second respondent may answer "yes" or "no" at will, with the only restriction being that the second answer must not logically contradict the first. Each successive reply is made so it is consistent with the previous clues, and the game ends when the inquirer asks: "Is the word _____?," and the respondent is forced to reply "yes." The game is played, and without further discussion, the guests sit down to dinner. To the seeker, and to any guest who arrived in the middle of the game, there was no reason to doubt that a chosen object word (goal) existed from the outset of the questioning. To him, the only mystery is why so many knowing looks were exchanged during dinner!

Cohen (1976) described Polya's urn and other dynamically evolving situations similar to Wheeler's parlor game, where if the "game" is repeated, the results will be quite different. Wheeler speculates that his parlor game might have a lot in common with the development of science, laying open the question of whether the corpus of science is really the unique structure most consider it to be. ·

Wheeler's parlor game, Polya's urn, or other similar procedures are defined by an unchanging set of rules. However, each realization of the "game" transpires and ends uniquely. A variational principle plays a role analogous to such a set of rules. Given the present state of a system (and, in general, its past history), the variational rule describes the *direction* in which the system should make the next infinitesimal progression. Once it has advanced, the configuration has altered, and external influences have changed in the interim. It is no longer exactly the same system, and the unchanging variational rule now points out a new direction in which to make the next step. If the system achieves a local optimum (the rule says no further change), there is no way of knowing from a single realization whether the final state existed a priori or not. It is possible that repetition of the "experiment" or "game" could show that the end-state is not unique or is possibly distributed in a uniform or random manner (Cohen, 1976). Hence, it is wrong to necessarily equate variational rules with a priori goal-seeking or with teleology.

To the best of anyone's knowledge, the evolution of the biosphere is a solitary event. Barring the discovery of *several* independent extraterrestrial realms of life, it remains impossible to make any scientific statement concerning the existence or nonexistence of goals for evolution. Yet this issue so impacts the human psyche that intelligent individuals without opinions on this metaphysical question are rare indeed. Furthermore, it is extremely difficult, and probably not desirable, to keep such inclinations from motivating an individual's scientific pursuits. To some degree, like it or not, everyone's observations are influenced by subjective beliefs. What must be scrupulously avoided by the phenomenologist, however, is any statement about or description of natural phenomena that purports to "prove" or unambiguously resolve the answer to metaphysical questions. In fact, one might argue that a good phenomenological description portrays natural events in as philosophically neutral a vein as possible.

These epistemological considerations are mentioned, because the principle of maximal ascendency has been judged "too transcendental" or "dream-like" by some of this author's friends and critics. Does not this author's emphasis on influence from larger scale events and the rejection of the strictly reductionistic perspective reveal a bias that many would consider metaphysical?

However, it could be argued as easily that prohibiting top-down influence is itself a metaphysical assertion. Phenomenological science should quantitatively portray natural events in a manner that can be verified by any observers, regardless of their opinions and beliefs. How neutral is it to limit description to unidirectional causality when allowing for the possibility of influence from any scale requires less presupposition? Of course, to some, this equanimity opens Pandora's box. Once the possibility of top down causality is admitted, where will it all end? However, the same question could be applied to the lower end of the hierarchy, as news reports almost daily herald the discovery of new entities that are components of "atoms" earlier thought to be indivisible.

In the end, what is most important is that phenomenological science be restricted to the realm of the observable. While new tools may expand this realm in the future, society is not about to perceive the ultimate anytime soon (barring, that is, the insanity of thermonuclear catastrophe!)

If all parties agree to restrict discussion to the perceptible, then the principle of optimal ascendency can be reconciled to most philosophies. In ecology, for example, the description disregards the actual mechanisms working to increase community ascendency. Darwinian selection has been suggested as a class of mechanisms that define maturing networks (Section 4.2). However, critics of Darwin (Ho and Saunders, 1979) could point out that their mechanisms also increase ascendency. Absurdists such as Monod could highlight the role that chance plays in development (ascendency is built upon probability theory) and could argue that the instantaneous apparent goal of the cosmos is meaningless in light of the infinity of possible alternatives. None of this prohibits transcendentalists such as Theilhard de Chardin from seeing the rise of ascendency as similar to a Hegelian progression from matter to spirit—the universe attracted towards its ultimate omega point.

However, if ascendency cannot help one to judge between opposing philosophies, what good is it? Here it should be recalled that the aim of phenomenological science is not to explain, but to describe. The world itself admits to description by many conflicting philosophies. So it should also be with faithful descriptions of nature—and ascendency is intended to be faithful to the point of approaching a *quantitative* tautology. The means are now at hand to assign a number to any given configuration of processes. Numerical comparisons of organization can now be made, where before one had to be satisfied with only vague subjective notions. The course of an economy or an ecosystem may now be charted using the information variables. The impacts of natural events or anthropogenic changes upon the whole system may be measured and evaluated. Perhaps most importantly, the principle of optimal ascendency provides a unifying framework for diverse and sometimes contradictory trends observed in the natural and social realms. To say these activities do not belong to the realm of science, one must also abrogate thermodynamics and intentionally blind oneself to the description of fundamental order in the universe.

There is yet another role for the idea of maximal ascendency. At a recent meeting devoted to the evolving discipline of "cognitive science," Professor Peter M. Allen of the Free University at Brussels prefaced a few remarks to the group with his concern about the "human condition." He was referring primarily to the psychic unease that results from contemplating oneself, humankind, and the natural world. Now, reductionistic science has served admirably to advance the material condition of humankind. Everyone is grateful for the degree of freedom from disease and hunger wrought by discoveries in microbiology and molecular biology. However, the attendant reductionistic attitudes about nature are anything but ennobling to the human self-image. If all cause originates in the netherworld of molecules and beyond, what significance does an individual have? Is it a person that acts to some end, or just a big collection of molecules? Actually, classical thermodynamics offers little more by way of encouragement. Things are either conserved, or they run down. In all, most of existing science portrays the universe as a rather dismal place.

That chaos abounds, hardly anyone would deny. However, a universe of nothing but chaos contradicts common experience every bit as much as Laplace's divining angel. Organization does exist amidst the confusion, and it is often seen to spread. As the scale of observation increases, new organizations appear with behaviors that in a real sense are autonomous of any description at finer resolution. Over the course of a few weeks, changes in an ecosystem are highly influenced by the DNA strands possessed by the community. Yet over thousands of years, the genetic makeup of a biome is shaped by the larger environment *and* by the history of ecosystems configurations. Over weeks or months, an individual may lose his mental faculties because of organic disturbance; but over a lifetime, that same mind is seen to have been a true agent for change in the world.

The principle of optimal ascendency is an attempt to give quantitative meaning to this current that runs against (but coexists with) the rush towards chaos, an effort to highlight the possibility of influence directed from living beings towards the lesser world, and an endeavor to reconcile scientific discourse with the fact that living things grow and develop.

Review of Matrix and Vector Operations

For the reader who may be unfamiliar with elementary matrix and vector operations, or for those who wish to refresh these skills, a very brief presentation of those elements of matrix algebra used in this book may prove helpful. What follows is intended to be neither complete nor rigorous, and anyone seeking a more thorough foundation in this subject is urged to consult any standard text on the subject.

A two-dimensional matrix is a collection of mathematical elements, or components, arranged in rectangular fashion by rows and columns. Matrices are denoted by capital letters enclosed in square brackets. For example, [A] and [B] might denote the following 3×3 arrays of integers:

$$[A] = \begin{bmatrix} 3 & 6 & 8 \\ 7 & 4 & 1 \\ 2 & 9 & 5 \end{bmatrix}, \tag{A.1}$$

$$[B] = \begin{bmatrix} 1 & 3 & 1 \\ 3 & 2 & 5 \\ 2 & 1 & 4 \end{bmatrix}. \tag{A.2}$$

The elements of an array are identified by subscripts denoting the row and column, respectively, in which they appear. Thus, $A_{32} = 9$ and $B_{31} = 2$.

The addition of two matrices is possible only when they possess the same numbers of rows and columns. In such cases, the sum of two matrices is formed simply by adding the respective elements of the summand matrices. Hence,

$$[C] = [A] + [B] \tag{A.3}$$

implies

$$C_{ij} = A_{ij} + B_{ij} \text{ for all } i, j. \tag{A.4}$$

If [A] and [B] are the matrices identified in (A.1–2), then

$$[C] = \begin{bmatrix} 4 & 9 & 9 \\ 10 & 6 & 6 \\ 4 & 10 & 9 \end{bmatrix}.$$

(A.5)

Similarly, the difference of two matrices is formed by taking the differences between the corresponding components,

$$D_{ij} = A_{ij} - B_{ij} \text{ for all } i,j,$$

(A.6)

so that for (A.1–2)

$$[D] = \begin{bmatrix} 2 & 3 & 7 \\ 4 & 2 & -4 \\ 0 & 8 & 1 \end{bmatrix}.$$

(A.7)

The product of two matrices is formed from the components of its factors in a fashion not resembling the addition or subtraction of matrices. Before the product of two matrices can be taken, it is necessary that the number of columns in the multiplier matrix (the one on the left) be equal to the number of rows in the multiplicand matrix. The product matrix will possess the same number of rows as does the multiplier, and its number of columns is identical to that of the multiplicand. For example, a 2×3 matrix may multiply a 3×4 matrix, because the number of columns in the first corresponds to the number of rows in the second. The resulting product will have 2 rows and 4 columns.

The i-jth element of the product matrix is formed from the ith row of the multiplier and the jth column of the multiplicand. Because the number of elements in these two strings of components is equal, the corresponding elements of the two sequences may be multiplied, and all such products may then be summed to give the desired element of the product matrix.

Supposing that

$$[E] = \begin{bmatrix} 2 & 1 & 4 \\ 3 & 2 & 5 \end{bmatrix},$$

(A.8)

and

$$[F] = \begin{bmatrix} 1 & 0 & 1 & 0 \\ 3 & 2 & 0 & 1 \\ 5 & 6 & 8 & 0 \end{bmatrix},$$

(A.9)

then to calculate the 2–3 component of the product matrix $[P] = [E] [F]$, one singles out the second row of $[E]$ and the third column of $[F]$. The corresponding elements of these two sequences are multiplied, and the products are accumulated:

$$P_{23} = E_{21}F_{13} + E_{22}F_{23} + E_{23}F_{33},$$

or

$$P_{23} = 3 \times 1 + 2 \times 0 + 5 \times 8 = 43. \tag{A.10}$$

It is helpful to envision this process as placing a finger of the left hand on the first element of the second row of $[E]$ and a finger of the right hand on the first entry in the third column of $[F]$. These numbers (3 and 1) are multiplied and the product noted. The fingers are then advanced to the next members of the row and column (2 and 0, respectively), and the product of that pair is added to the first. Finally, the pointers advance to the third pair (5 and 8) and that product is added to the previous accumulation. The reader may practice forming the remaining elements of $[P]$ as follows:

$$[P] = [E][F] = \begin{bmatrix} 2 & 1 & 4 \\ 3 & 2 & 5 \end{bmatrix} \begin{bmatrix} 1 & 0 & 1 & 0 \\ 3 & 2 & 0 & 1 \\ 5 & 6 & 8 & 0 \end{bmatrix} = \begin{bmatrix} 25 & 26 & 34 & 1 \\ 34 & 34 & 43 & 2 \end{bmatrix}. \tag{A.11}$$

This process may be generalized to the product of any $l \times m$ matrix, $[G]$, multiplying any $m \times n$ matrix, $[H]$, as

$$P_{ij} = \sum_{k=1}^{m} G_{ik}H_{kj} \qquad i = 1, 2, \ldots l; \qquad j = 1, 2, \ldots n. \tag{A.12}$$

A square matrix, such as $[A]$ or $[B]$, is one possessing an identical number of rows and columns. It is possible to multiply two square matrices, either as $[A][B]$ or $[B][A]$. Care should be exercised, however, to specify the order of multiplication, because in general the results will not be the same; that is, matrix multiplication is not always commutative:

$$[A][B] \neq [B][A]. \tag{A.13}$$

Integer powers of a square matrix are calculated by successive multiplication of the base matrix; namely,

$$[B]^2 = \begin{bmatrix} 1 & 3 & 1 \\ 3 & 2 & 5 \\ 2 & 1 & 4 \end{bmatrix} \begin{bmatrix} 1 & 3 & 1 \\ 3 & 2 & 5 \\ 2 & 1 & 4 \end{bmatrix} = \begin{bmatrix} 12 & 10 & 20 \\ 19 & 18 & 33 \\ 13 & 12 & 23 \end{bmatrix}. \tag{A.14}$$

Of particular importance is the identity matrix, $[I]$, with elements $\delta_{ij} = 1$ for $i = j$ and $\delta_{ij} = 0$ for $i \neq j$ (i.e, 1's along the diagonal and zeroes elsewhere). Multiplying any matrix from either side by an identity matrix with the appropriate number of rows or columns will regenerate the starting matrix:

$$[I][A] = [A] \tag{A.15}$$

or

$$\begin{bmatrix} 1 & 0 & 0 \\ 0 & 1 & 0 \\ 0 & 0 & 1 \end{bmatrix} \begin{bmatrix} 3 & 6 & 8 \\ 7 & 4 & 1 \\ 2 & 9 & 5 \end{bmatrix} = \begin{bmatrix} 3 & 6 & 8 \\ 7 & 4 & 1 \\ 2 & 9 & 5 \end{bmatrix}. \tag{A.16}$$

Also, $[A][I] = [A]$. Interjecting the identity matrix as a factor into any equation does not affect the relationship. For example,

$$[A][B] = [A][I][B]. \tag{A.17}$$

Having learned to add, subtract, and multiply matrices, the reader might also inquire about the process of division. Strictly speaking, one does not divide one matrix by another, but the same end may be achieved by multiplying the first by the inverse of the second. The inverse of a square matrix is another square matrix of the same dimension, which when multiplied by the given matrix (from either side) yields the identity matrix. The inverse of a matrix will be denoted by posing the superscript -1 after the given matrix. Thus, the inverse of $[B]$ is $[B]^{-1}$, and

$$[B]^{-1}[B] = [B][B]^{-1} = [I]. \tag{A.18}$$

Not every matrix possesses a unique inverse, and matrices without proper inverses are called singular. Singular matrices are rarely encountered in the analyses covered in this text.

How to calculate the inverse of a given matrix will not be discussed in detail here; suffice it to say that one may regard (A.18) as a set of n^2 simultaneous linear equations with the components of $[B]^{-1}$ as the unknowns. Solving (A.18), one finds that

$$\begin{bmatrix} -.75 & 2.75 & -3.25 \\ .50 & -.50 & .50 \\ .25 & -.25 & 1.75 \end{bmatrix} \begin{bmatrix} 1 & 3 & 1 \\ 3 & 2 & 5 \\ 2 & 1 & 4 \end{bmatrix} = \begin{bmatrix} 1 & 0 & 0 \\ 0 & 1 & 0 \\ 0 & 0 & 1 \end{bmatrix}. \tag{A.19}$$

The final matrix operation to be covered is that of transposition. To transpose a matrix, one interchanges its rows and columns. The result is denoted by a superscript T after the original matrix. For example,

$$[A]^T = \begin{bmatrix} 3 & 7 & 2 \\ 6 & 4 & 9 \\ 8 & 1 & 5 \end{bmatrix}. \tag{A.20}$$

The reader may verify that in general,

$$[[A]^T]^T = [A], \tag{A.21}$$

and

$$\{[A][B]\}^T = [B]^T[A]^T. \qquad (A.22)$$

For nonsingular square matrices

$$\{[A]^{-1}\}^T = \{[A]^T\}^{-1}. \qquad (A.23)$$

In this book, all vectors are treated as single columns of elements and are denoted by capital letters enclosed in parentheses; for example,

$$(U) = \begin{pmatrix} 9 \\ 1 \\ 7 \end{pmatrix} \qquad (A.24)$$

$$(V) = \begin{pmatrix} 2 \\ 1 \\ 2 \end{pmatrix}. \qquad (A.25)$$

As with matrices, adding and subtracting vectors is simply a matter of performing these operations on the corresponding components of the vectors being combined. For example,

$$W_i = U_i + V_i, \qquad i = 1, 2, \ldots n, \qquad (A.26)$$

or

$$(W) = (U) + (V) = \begin{pmatrix} 9 \\ 1 \\ 7 \end{pmatrix} + \begin{pmatrix} 2 \\ 1 \\ 2 \end{pmatrix} = \begin{pmatrix} 11 \\ 2 \\ 9 \end{pmatrix}, \qquad (A.27)$$

and

$$S_i = U_i - V_i, \qquad i = 1, 2, \ldots, n, \qquad (A.28)$$

or

$$(S) = (U) - (V) = \begin{pmatrix} 9 \\ 1 \\ 7 \end{pmatrix} - \begin{pmatrix} 2 \\ 1 \\ 2 \end{pmatrix} = \begin{pmatrix} 7 \\ 0 \\ 5 \end{pmatrix}. \qquad (A.29)$$

The last of the operations used in this text is the multiplication of a vector by a matrix. To perform this operation, the number of columns in the matrix must equal the number of components in the vector. The operation yields a vector with the same number of elements as there are rows in the matrix. The actual multiplication proceeds exactly as if one were computing any single row in the product of two matrices. In terms of components:

$$Q_i = \sum_{j=1}^{n} E_{ij} V_j,$$ (A.30)

or

$$(Q) = [E](V) = \begin{bmatrix} 2 & 1 & 4 \\ 3 & 2 & 5 \end{bmatrix} \begin{pmatrix} 2 \\ 1 \\ 2 \end{pmatrix} = \begin{pmatrix} 13 \\ 18 \end{pmatrix}.$$ (A.31)

As with matrix multiplication, the multiplication of a matrix by a vector is associative (but obviously not commutative); namely,

$$[A]\{(U) + (V)\} = [A](U) + [A](V),$$ (A.32)

and

$$\{[A] + [B]\}(U) = [A](U) + [B](U).$$ (A.33)

The identity matrix plays the same role with vectors as it does with matrices, that is,

$$[I](V) = (V).$$ (A.34)

Once these individual operations have been mastered, they may be used in sequence as an algebra for the purpose of solving matrix and vector equations. As an example, one wishes to solve the following matrix-vector system of equations for the vector (T):

$$(T) = [G](T) + (R).$$ (A.35)

One begins by transposing all members containing (T) to the left-hand side of the equation

$$(T) - [G](T) = (R).$$ (A.36)

The object is to factor (T) out of the left-hand side. But a matrix multiplier for the first (T) seems to be missing. The difficulty is only apparent because one may interject the identity matrix as a factor at any time (A.34),

$$[I](T) - [G](T) = (R).$$ (A.37)

Out of which, by associativity (A.3), the above becomes

$$\{[I] - [G]\}(T) = (R).$$ (A.38)

The multiplier of (T) on the left-hand side must be eliminated. In conventional algebra, one divides both sides of the equation by the multiplier of the unknown variable. The equivalent operation in matrix algebra is to multiply both sides by the inverse of the multiplier matrix, assuming such an inverse exists. Hence,

$$\{[I] - [G]\}^{-1}\{[I] - [G]\}(T) = \{[I] - [G]\}^{-1}(R),$$ (A.39)

which according to the definition of an inverse (A.18) is the same as,

$$[I](T) = \{[I] - [G]\}^{-1}(R), \qquad\qquad (A.40)$$

or, applying (A.34) a second time, the desired solution,

$$(T) = \{[I] - [G]\}^{-1}(R), \qquad\qquad (A.41)$$

is obtained.

With a minimum of practice, anyone with basic skills in conventional algebra can become proficient in the analogous matrix operations.

A Program to Calculate Information Indices

The manual calculation of network information indices (ascendency, redundancy, etc.) for any but the simplest of networks is a very tedious task. In order to work the examples given in this text and to allow the reader to explore how these indices behave as networks change, it is advisable that the reader employ computational machinery (Emery, personal communication). As an aid to those wishing to calculate these indices on a microcomputer, the BASIC program listed below can be used to compute all the relevant variables.

This program has been designed to run on practically any BASIC compiler. Only those elementary functions, which do not vary significantly from machine to machine, are used. Operations are broken down into simple steps, and only single character variable names have been used. These constraints cause the program to be long and cryptic, but almost universally applicable.

Nevertheless, a few characteristics of the program may still differ among compilers. Here the input data are taken from the Cone Spring flow network (Figure 3.2) and are inserted into the program as data statements in lines 150, 450, 500, 550, and 620 to 660. Very elementary compilers may require some other format. The function "LOG" first appearing in line 810 is intended to compute natural logarithms to the base e. On some compilers this function is designated by "LN." Similarly, the single quotation marks appearing first in the PRINT statement of line 1230 may become double marks on some machines. Those using slow microcomputers may become anxious waiting for the first output to appear. By changing all the REM statements to PRINT, one may remain apprised of which part of the program is being executed at reasonably brief intervals. Expert programmers are encouraged to abbreviate the program to suit their own needs.

Program Listing

```
100 DIM Q(20)
110 DIM U(20)
120 DIM P(20,20)
130 REM READ IN THE NUMBER OF COMPONENTS.
140 READ N
150 DATA 5
160 LET K=N+1
170 LET L=N+2
180 LET M=N+3
190 REM ZERO ALL VARIABLES.
200 LET A=0.
210 LET B=0.
220 LET C=0.
230 LET D=0.
240 LET E=0.
250 LET F=0.
260 LET G=0.
270 LET R=0.
280 LET S=0.
290 LET T=0.
300 LET W=0.
310 LET Y=0.
320 LET Y=0.
330 LET Z=0.
340 FOR I=1 TO M
350 LET Q(I)=0.
360 LET U(I)=0.
370 FOR J=1 TO M
380 LET P(I,J)=0.
390 NEXT J
400 NEXT I
410 REM READ IN VECTOR OF EXTERNAL INPUTS
420 FOR I=1 TO N
430 READ P(K,I)
440 NEXT I
450 DATA 11184.,635.,0.,0.,0.
460 REM READ IN VECTOR OF USEFUL OUTPUTS.
470 FOR I=1 TO N
480 READ P(I,L)
490 NEXT I
500 DATA 300.,860.,255.,0.,0.
510 REM READ IN VECTOR OF RESPIRATIONS.
520 FOR I=1 TO N
```

```
530 READ P(I,M)
540 NEXT I
550 DATA 2003.,3109.,3275.,1814.,203.
560 REM READ IN THE N ROWS OF THE EXCHANGE MATRIX.
570 FOR I=1 TO N
580 FOR J=1 TO N
590 READ P(I,J)
600 NEXT J
610 NEXT I
620 DATA 0.,8881.,0.,0.,0.
630 DATA 0.,0.,5205.,2309.,0.
640 DATA 0.,1600.,0.,75.,0.
650 DATA 0.,200.,0.,0.,370.
660 DATA 0.,167.,0.,0.,0.
670 REM CALCULATE TST AND COMPARTMENTAL SUMS.
680 FOR I=1 TO M
690 FOR J=1 TO M
700 LET Q(I)=Q(I)+P(I,J)
710 LET U(I)=U(I)+P(J,I)
720 NEXT J
730 LET T=T+Q(I)
740 NEXT I
750 REM CALCULATE ASCENDENCIES BY COMPONENTS.
760 FOR I=1 TO M
770 FOR J=1 TO M
780 If P(I,J)<=0. THEN GO TO 820
790 LET G=(P(I,J)*T)/(Q(I)*U(J))
800 LET A=A+P(I,J)*LOG(G)
810 IF I<=N AND J<=N THEN LET B=B+P(I,J)*LOG(G)
820 NEXT J
830 NEXT I
840 REM CALCULATE OTHER INFORMATION COMPONENTS.
850 LET G=Q(K)/T
860 IF Q(K)>0. THEN LET D=-Q(K)*LOG(G)
870 LET C=D
880 FOR I=1 TO N
890 IF P(I,L)<=0. THEN GO TO 940
900 LET G=P(I,L)/U(L)
910 LET W=W-P(I,L)*LOG(G)
920 LET G=Q(I)/T
930 LET E=E-P(I,L)*LOG(G)
940 IF P(I,M)<=0. THEN GO TO 990
950 LET G=P(I,M)/U(M)
960 LET Z=Z-P(I,M)*LOG(G)
970 LET G=Q(I)/T
```

```
980 LET S=S−P(I,M)*LOG(G)
990 LET G=Q(I)/T
1000 LET Q(I)>0. THEN LET C=C−Q(I)*LOG(G)
1010 IF P(K,I)<=0. THEN GO TO 1040
1020 LET G=P(K,I)/U(I)
1030 LET F=F−P(K,I)*LOG(G)
1040 FOR J=1 TO N
1050 IF P(I,J)<=0. THEN GO TO 1080
1060 LET G=P(I,J)/U(J)
1070 LET R=R−P(I,J)*LOG(G)
1080 NEXT J
1090 NEXT I
1100 REM CONVERT ALL INDICES FROM NATS TO BITS.
1110 LET X=1.442695
1120 LET A=X*A
1130 LET H=X*(C-D)
1140 LET C=X*C
1150 LET B=X*B
1160 LET R=X*R
1170 LET E=X*E
1180 LET S=X*S
1190 LET F=X*F
1200 LET W=X*W
1210 LET Z=X*Z
1220 REM WRITE OUT THE RESULTS.
1230 PRINT 'TOTAL SYSTEM THROUGHPUT IS',T
1240 PRINT ''
1250 PRINT 'FULL DEVELOPMENT CAPACITY IS',C
1260 LET Y=A/C
1270 PRINT 'FULL ASCENDENCY IS',A,Y
1280 LET Y=F/C
1290 PRINT 'OVERHEAD ON INPUTS IS',F,Y
1300 LET Y=W/C
1310 PRINT 'OVERHEAD ON EXPORTS IS',W,Y
1320 LET Y=Z/C
1330 PRINT 'OVERHEAD ON RESPIRATIONS IS',Z,Y
1340 LET Y=R/C
1350 PRINT 'SYSTEM REDUNDANCY IS',R,Y
1360 PRINT ''
1370 PRINT 'INTERNAL CAPACITY IS',H
1380 LET Y=B/H
1390 PRINT 'INTERNAL ASCENDENCY IS',B,Y
1400 LET Y=E/H
1410 PRINT 'TRIBUTE TO OTHER SYSTEMS IS ',E,Y
1420 LET Y=S/H
```

1430 PRINT 'DISSIPATION IS ',S,Y
1440 LET Y=R/H
1450 PRINT 'SYSTEM REDUNDANCY IS',R,Y
1460 END

Upon execution, the user should see something resembling the output below. The results listed in the first column of numbers have the dimensions kcal bits m^{-2} y^{-1} (except for the total throughput, which is in kcal m^{-2} y^{-1}). The numbers in the second column indicate the fractional amounts of the relevant capacity comprised by each component. For example, the full ascendency constitutes about 60.9% of the full capacity, while the tribute makes up only 4.2% of the internal capacity. The value of any information index in bits is obtained by dividing the appropriate scaled quantity by the total systems throughput.

Sample Output

TOTAL SYSTEM THROUGHPUT IS	42445	
FULL DEVELOPMENT CAPACITY IS	93171.564	
FULL ASCENDENCY IS	56725.481	.60882825
OVERHEAD ON INPUTS IS	2652.1404	.02846513
OVERHEAD ON EXPORTS IS	1919.568	.02060251
OVERHEAD ON RESPIRATIONS IS	21364.052	.22929799
SYSTEM REDUNDANCY IS	10510.32	.11280609
INTERNAL CAPACITY IS	71371.577	
INTERNAL ASCENDENCY IS	29331.977	.41097561
TRIBUTE TO OTHER SYSTEMS IS	2971.3334	.04163189
DISSIPATION IS	28557.946	.40013051
SYSTEM REDUNDANCY IS	10510.32	.14726198

Other programs (written in ANSI FORTRAN), which perform most of the calculations appearing in any of the examples throughout this book, are available from the author upon request.

References

Abramson, N. 1963. Information Theory and Coding. McGraw-Hill, New York. 201 p.

Aczel, J. and Z. Daroczy. 1975. On Measures of Information and Their Chracterizations. Academic Press, New York. 234 p.

Allen, T. F. H., and T. B. Starr. 1982. Hierarchy. University of Chicago Press, Chicago, 310 p.

Amir, S. 1979. Economic interpretations of equilibrium concepts in ecological systems. J. Social. Biol. Struct. 2:293–314.

Amir, S. 1983. Ecosystem productivity and persistence: on the need for two complementary views in evaluating ecosystem functioning. Ecosystem Research Center Report ERC-047, Cornell University, Ithaca, New York. 103 p.

Andresen, B., P. Salamon, and R. S. Berry. 1977. Thermodynamics in finite time: extremals for imperfect heat engines. J. Chem. Phys. 66:1571–1577.

Atlan, H. 1974. On a formal definition of organization. J. theor. Biol. 45:295–304.

Augustinovics, M. 1970. Methods of international and intertemporal comparison of structure. In: A. P. Carter and A. Brody [eds.], Contributions to Input-Output Analysis, Vol. I, North Holland, Amsterdam. 345 p., pp. 249–269.

Ayres, R. U., and A. V. Kneese. 1969. Production, consumption and externalities. Am. Econ. Rev. 59(3):282–287.

Bird, R. B., W. E. Stewart, and E. N. Lightfoot. 1960. Transport Phenomena. John Wiley and Sons, New York. 780 p.

Boltzmann, L. 1872. Weitere Studien über das Wärmegleichgewicht unter Gasmolekülen. Wien. Ber. 66:275–370.

Bormann, F. H. and G. E. Liken. 1979. Pattern and Process in a Forested Ecosystem. Springer-Verlag, New York. 253 p.

Bosserman, R. W. 1981. Sensitivity techniques for examination of input-output flow analyses. In: W. J. Mitsch and J. M. Klopatek [eds.], Energy and Ecological Modelling, Elsevier, Amsterdam. 839 p., pp. 653–660.

Boulding, K. E. 1978. Ecodynamics: A New Theory of Societal Evolution. Sage Publications, Beverly Hills, California. 368 p.

Boulding, K. E. 1982. The unimportance of energy. In: W. J. Mitsch, R. K. Ragade, R. W. Bosserman and J. A. Dillon, Jr. [eds.], Energetics and Systems, Ann Arbor Science, Ann Arbor, Michigan, 132 p., pp. 101–108.

Brillouin, L. 1956. Science and Information Theory. Academic Press, New York. 320 p.

Caratheodory, C. 1909. Untersuchungen über die Grundlagen der Thermodynamik. Math. Ann. 67:355–386.

Carnot, S. 1824. Reflections on the Motive Power of Heat (translated 1943). ASME, New York. 107 p.

Chapman, S., and T. G. Cowling. 1961. The Mathematical Theory of Non-Uniform Gases. Cambridge University Press, London. 431 p.

Cheslak, E. F., and V. A. Lamarra. 1981. The residence time of energy as a measure of ecological organization. In: W. J. Mitsch and R. W. Bosserman [eds.], Energy and Ecological Modelling, Elsevier, New York. 839 p., pp. 591–600.

Cheung, A. K-T. 1985a. Network Optimization in Ecosystem Development. Doctoral Dissertation, Department of Mathematical Sciences, The Johns Hopkins University, Baltimore, Maryland. 163 p.

Cheung, A. K-T. 1985b: ECONET: Algorithms for network optimization in ecosystem development analysis. Technical Report No. 423, Department of Mathematical Sciences, The Johns Hopkins University, Baltimore, Maryland. 63 p.

Clements, F. E., and V. E. Shelford. 1939. Bio-ecology. John Wiley and Sons, New York. 425 p.

Cohen, J. E. 1976. Irreproducible results and the breeding of pigs. Bioscience 26:391–394.

Conrad, M. 1972. Statistical and hierarchical aspects of biological organization. In: C. H. Waddington [ed.], Towards a Theoretical Biology, Vol. 4, University Edinburgh Press, Edinburgh. 299 p., pp. 189–220.

Conrad, M. 1983. Adaptability: The Significance of Variability from Molecule to Ecosystem. Plenum Press, New York. 383 p.

Corning, P. A. 1983. The Synergism Hypothesis: A Theory of Progressive Evolution. McGraw-Hill, New York. 492 p.

Costanza, R., and C. Neill. 1984. Energy intensities, interdependence, and value in ecological systems: a linear programming approach. J. theor. Biol. 106:41–57.

Daly, H. E. 1968. On economics as a life science. J. Political Econ. 76:392–405.

Dawkins, R. 1976. The Selfish Gene. Oxford University Press, New York. 224 p.

Eigen, M. 1971. Selforganization [sic] of matter and the evolution of biological macromolecules. Naturwiss 58:465–523.

Elder, D. 1979. Why is regenerative capacity restricted in higher organisms? J. theor. Biol. 81:563–568.

Elsasser, W. M. 1981. Principles of a new biological theory: a summary. J. theor. Biol. 89:131–150.

Engelberg, J., and L. L. Boyarsky. 1979. The noncybernetic nature of ecosystems. Am. Nat. 114:317–324.

Finn, J. T. 1976. Measures of ecosystem structure and function derived from analysis of flows. J. theor. Biol. 56:363–380.

Finn, J. T. 1980. Flow analysis of models of the Hubbard Brook ecosystem. Ecology 61:562–571.

Fontaine, T. D. 1981. A self-designing model for testing hypotheses of ecosystem development. In: S. E. Jørgensen [ed.], Progress in Ecological Engineering and Management by Mathematical Modelling, Elsevier, Amsterdam. 1014 p., pp. 281–291.

Georgescu-Roegen, N. 1971. The Entropy Law and the Economic Process. Harvard University Press, Cambridge, Massachusetts. 457 p.

Gladyshev, G. P. 1982. Classical thermodynamics, tandemism and biological evolution. J. theor. Biol. 94:225–239.

Glansdorff, P., and I. Prigogine. 1971. Thermodynamic Theory of Structure, Stability and Fluctuations. Wiley-Interscience, London. 306 p.

Gleason, H. A. 1926. The individualistic concept of the plant association. Bull. Torrey Bot. Club 53:1–20.

Goel, N. S., S. C. Maitra, and E. W. Montroll. 1971. On the Volterra and Other Nonlinear Models of Interacting Populations. Academic Press, New York. 145 p.

Goldstein, H. 1950. Classical Mechanics. Addison-Wesley, Cambridge, Massachusetts. 399 p.

Halfon, F A, 1979. Theoretical Systems Ecology. Academic Press, New York. 516 p.

Hannon, B. 1973. The structure of ecosystems. J. theor Biol, 41:535–546.

Hannon, B. 1979. Total energy costs in ecosystems. J. theor. Biol. 80:271–293.

Hawkins, D., and H. A. Simon. 1949. Note: some conditions of macroeconomic stability. Econometrica 17:245–248.

Hill, J. IV and R. G. Wiegert. 1980. Microcosms in ecological modeling. In: J. P. Giesy [ed.], Microcosms in Ecological Research. U.S. Department of Energy, Springfield, VA. 1110 p., pp. 138–163.

Hippe, P. W. 1983. Environ analysis of linear compartmental systems: the dynamic, time-invariant case. Ecol. Modelling. 19:1–26.

Hirata, H., and R. E. Ulanowicz. 1984. Information theoretical analysis of ecological networks. Int. J. Systems Sci. 15:261–270.

Ho, M. W., and P. T. Saunders. 1979. Beyond Neo-Darwinism—an epigenetic approach to evolution. J. theor. Biol. 78:573–591.

Hutchinson, G. E. 1948. Circular causal systems in ecology. Ann. N.Y. Acad. Sci. 50:221–246.

Isard, W. 1968. Some notes on the linkage of ecologic and economic systems. In: J. B. Parr and W. Isard [eds.], Regional Science Association: Papers XXII. University of Illinois Press, Urbana-Champaign, IL. 220 p., pp. 85–96.

Isard, W., C. L. Choguill, J. Kissin, R. H. Seyfarth, and R. Tatlock. 1972. Ecologic-Economic Analysis for Regional Development. Free Press, New York. 270 p.

Jaynes, E. T. 1979. Where do we stand on maximum entropy? In: R. Levine and M. Tribus [eds.], The Maximum Entropy Formalism, MIT Press, Cambridge, Massachusetts. 498 p., pp. 15–118.

Johnson, L. 1981. The thermodynamic origin of ecosystems. Can. J. Fish. Aquat. Sci. 38:571–590.

Jørgensen, S. E., and H. Mejer. 1979. A holistic approach to ecological modelling. Ecol. Modelling 7:169–189.

Jørgensen, S. E., and H. Mejer. 1981. Exergy as a key function in ecological models. In: W. J. Mitsch and R. W. Bosserman [eds.], Energy and Ecological Modelling, Elsevier, New York, 839 p., pp. 587–590.

Katchalsky, A., and P. Curran. 1965. Non-Equilibrium Thermodynamics in Biophysics. Harvard University Press, Cambridge, Massachusetts. 248 p.

Kennington, J. L., and R. V. Helgason. 1980. Algorithms for Network Programming. John Wiley and Sons, New York. 291 p.

Kerner, E. H. 1957. A statistical mechanics of interacting biological species. Bull. Math. Biophys. 19:121–146.

Knuth, D. E. 1973. Fundamental Algorithms, Vol. 1. Addison-Wesley, Reading, Massachusetts. 228 p.

Kubat, L., and J. Zeman. 1975. Entropy and Information in Science and Philosophy. Elsevier, Amsterdam. 260 p.

Kullback, S. 1959. Information Theory and Statistics. Peter Smith, Gloucester, Massachusetts. 399 p.

Lange, O. 1963. Political Economy. Pergamon Press, New York. 355 p.

Laplace, P. S. 1814. A Philosophical Essay on Probabilities (translation by F. W. Truscott and F. L. Emory, 1951). Dover Publications, Inc., New York. 196 p.

Leontief, W. 1951. The Structure of the American Economy, 1919–1939, 2nd ed. Oxford University Press, New York. 257 p.

Leopold, L. B., and W. B. Langbein. 1966. River meanders. Sci. Am. 214(6):60–70.

Levine, S. 1980. Several measures of trophic structure applicable to complex food webs. J. theor. Biol. 83:195–207.

Lewin, R. 1984. Why is development so illogical? Science 224:1327–1329.

Lindeman, R. L. 1942. The trophic-dynamic aspect of ecology. Ecology 23:399–418.

Lorenz, E. N. 1963. Deterministic nonperiodic flow. J. Atmos. Sci. 20:130–141.

Lotka, A. J. 1922. Contribution to the energetics of evolution. Proc. Nat. Acad. Sci. 8:147–150.

Lurie, D., and J. Wagensberg. 1979. Non-equilibrium thermodynamics and biological growth and development. J. theor. Biol. 78:241–250.

MacArthur, R. H. 1955. Fluctuations of animal populations and a measure of community stability. Ecology 36:533–536.

MacMahon, J. 1979. Ecosystems over time: Succession and other type of changes. In: R. Waring [ed.], Forests: Fresh Perspectives from Ecosystem Analysis. Oregon State University Press, Corvallis. 199 p., pp. 27–58.

Margalef, R. 1968. Perspectives in Ecological Theory. University of Chicago Press, Chicago. 111 p.

Mateti, P., and N. Deo. 1976. On algorithms for enumerating all the circuits of a graph. SIAM J. Comput. 5:90–99.

May, R. M. 1973. Stability and Complexity in Model Ecosystems. Princeton University Press, Princeton, New Jersey. 235 p.

May, R. M. 1983. The structure of foodwebs. Nature 301:566–568.

Mayr, E. 1969. Principles of Systematic Zoology. McGraw-Hill, New York. 428 p.

McEliece, R. J. 1977. The Theory of Information and Coding. Addison-Wesley, Reading, Massachusetts. 302 p.

McGill, W. J. 1954. Multivariate information transmission. IRE Trans. Information Theory 4:93–111.

Odum, E. P. 1953. Fundamentals of Ecology. Saunders, Philadelphia. 384 p.

Odum, E. P., and H. T. Odum. 1959. Fundamentals of Ecology, 2nd ed. Saunders, Philadelphia. 546 p.

Odum, E. P. 1969. The strategy of ecosystem development. Science 164:262–270.

Odum, E. P. 1977. The emergence of ecology as a new integrative discipline. Science 195:1289–1293.

Odum, H. T., and R. C. Pinkerton. 1955. Time's speed regulator: the optimum efficiency for maximum power output in physical and biological systems. Am. Scientist 43:331–343.

Odum, H. T. 1971. Environment, Power and Society. John Wiley and Sons, New York. 331 p.

Onsager, L. 1931. Reciprocal relations in irreversible processes. Phys. Rev. 37:405–426.

Paltridge, G. W. 1975. Global dynamics and climate—a system of minimum entropy exchange. Qrt. J. R. Met. Soc. 101:475–484.

Patten, B. C., R. W. Bosserman, J. T. Finn, and W. G. Cale. 1976. Propagation of cause in ecosystems. In: B. C. Patten [ed.], Systems Analysis and Simulation in Ecology, Vol. 4, Academic Press, New York. 593 p., pp. 457–479.

Patten, B. C., and E. P. Odum. 1981. The cybernetic nature of ecosystems. Am. Nat. 118:886–895.

Patten, B. C. 1982. On the quantitative dominance of indirect effects in ecosystems. Unpublished paper presented at the Third International Conference on State-of-the-Art in Ecological Modeling, Colorado State University, May 24–28, Fort Collins, Colorado.

Patten, B. C. 1985. Energy cycling in the ecosystem. Ecol. Modelling 28:1–71.

Pimm, S. L., and J. H. Lawton. 1977. Number of trophic levels in ecological communities. Nature 268:329–331.

Pimm, S. L. 1982. Food Webs. Chapman and Hall, London. 219 p.

Platt, T. C., and K. Denman. 1975. Spectral analysis in ecology. Ann. Rev. Ecol. Syst. 6:189–210.

Powell, T. M., P. J. Richerson, T. M. Dillion, B. A. Agee, B. J. Dozier, D. A. Godden, and L. O. Myrup. 1975. Spatial scales of current speed and phytoplankton biomass fluctuations in Lake Tahoe. Science 189:1088–1090.

Preston, F. W. 1948. The commonness, and rarity of species. Ecology 29:254–283.

Prigogine, I. 1945. Moderation et transformations irreversibles des systemes ouverts. Bull. Classe Sci., Acad. Roy. Belg. 31:600–606.

Prigogine, I. 1947. Etude Thermodynamique des Phenomenes Irreversibles. Dunod, Paris. 143 p.

Prigogine, I. 1978. Time, structure and fluctuations. Science 201:777–785.

Prigogine, I. 1980. From Being to Becoming. W. H. Freeman, San Francisco. 272 p.

Prigogine, I., and I. Stengers. 1984. Order out of Chaos: Man's New Dialogue with Nature. Bantam, New York. 349 p.

Read, R. C., and R. E. Tarjan. 1975. Bounds on backtrack algorithms for listing cycles, paths, and spanning trees. Networks 5:237–252.

Richey, J. E., R. C. Wissmar, A. H. Devol, G. E. Likens, J. S. Eaton, R. G. Wetzel, W. E. Odum, N. M. Johnson, O. L. Loucks, R. T. Prentki, and P. H. Rich. 1978. Carbon flow in four lake ecosystems: a structural approach. Science 202:1183–1186.

Rutledge, R. W., B. L. Basorre, and R. J. Mulholland. 1976. Ecological stability: an information theory viewpoint. J. Theor. Biol. 57:355–371.

Scott, D. 1965. The determination and use of thermodynamic data in ecology. Ecology 46:673–680.

Shannon, C. E. 1948. A mathematical theory of communication. Bell System Tech.J. 27:379–423.

Simberloff, D. 1980. A succession of paradigms in ecology: essentialism to materialism and probabilism. Synthese 43:3–39.

Smerage, G. 1976. Matter and energy flows in biological and ecological systems. J. theor. Biol. 57:203–223.

Steele, J. H. 1974. The Structure of Marine Ecosystems. Harvard University Press, Cambridge, Massachusetts. 128 p.

Stent, G. 1981. Strength and weakness of the genetic approach to the development of the nervous system. Ann. Rev. Neurosci. 4:16–194.

Straskraba, M. 1983. Cybernetic formulation of control in ecosystems. Ecol. Modelling 18:85–98.

Tilly, L. J. 1968. The structure and dynamics of Cone Spring. Ecol. Monographs 38:169–197.

Tisza, L. 1966. Generalized Thermodynamics. MIT Press, Cambridge, Massachusetts. 384 p.

Tribus, M. 1961. Thermostatics and Thermodynamics. Van Nostrand, Princeton. 649 p.

Tribus, M., and E. C. McIrvine. 1971. Energy and information. Sci. Am. 225(3):179–188.

Ulanowicz, R. E. 1972. Mass and energy flow in closed ecosystems. J. theor. Biol. 34:239–253.

Ulanowicz, R. E. 1979. Prediction chaos and ecological perspective. In: E. A. Halfon [ed.], Theoretical Systems Ecology, Academic Press, New York. 516 p., pp. 107–117.

Ulanowicz, R. E., and W. M. Kemp. 1979. Toward canonical trophic aggregations. Am. Nat. 114:871–883.

Ulanowicz, R. E. 1980. An hypothesis on the development of natural communities. J. theor. Biol. 85:223–245.

Ulanowicz, R. E. 1983. Identifying the structure of cycling in ecosystems. Math. Biosci. 65:219–237.

Victor, P. A. 1972. Pollution: Economy and Environment. George Allen and Unwin, Ltd., London. 247 p.

Waddington, C. H. 1968. Towards a Theoretical Biology, Vol. 1. Edinburgh University Press, Edinburgh. 234 p.

Webster, J. R. 1979. Hierarchical organization of ecosystems. In: E. A. Halfon [ed.], Theoretical Systems Ecology, Academic Press, New York. 516 p., pp. 119–129.

Weiss, P. A. 1969. The living system: determinism stratified. In: A. Koestler and J. R. Smythies [eds.], Beyond Reductionism. MacMillan Co., New York. 438 p., pp. 3–55.

Westerhoff, H. V., K. J. Hellingwerf, and K. VanDam. 1983. Thermodynamic efficiency of microbial growth is low but optimal for maximal growth rate. Proc. Nat. Acad. Sci. 80:305–309.

Wicken, J. S. 1984. Autocatalytic cycling and self-organization in the ecology of evolution. Nature and System 6:119–135.

Wiener, N. 1948. Cybernetics. MIT Press, Cambridge, Massachusetts. 212 p.

Wills, G. 1978. Inventing America. Doubleday, Garden City, New York. 398 p.

Wilson, E. O. 1975. Sociobiology. Harvard University Press, Cambridge, Massachusetts. 697 p.

Woodwell, G. M., and Smith H. H. [eds.]. 1969. Diversity and Stability in Ecological Systems, Vol. 22, U.S. Brookhaven Symp. Biol., New York. 264 p.

Author Index

Subject Index